DOT COM MAN

Voices in Development Management

Series Editor:
Margaret Grieco
Napier University, Scotland

The Voices in Development Management series provides a forum in which grass roots organisations and development practitioners can voice their views and present their perspectives along with the conventional development experts. Many of the volumes in the series will contain explicit debates between various voices in development and permit the suite of neglected development issues such as gender and transport or the microcredit needs of low income communities to receive appropriate public and professional attention.

Also in the series

Dot Com Mantra
Social Computing in the Central Himalayas

PAYAL ARORA

Erasmus University Rotterdam, The Netherlands

Routledge
Taylor & Francis Group

LONDON AND NEW YORK

First published 2010 by Ashgate Publishing

2 Park Square, Milton Park, Abingdon, Oxon OX14 4RN
711 Third Avenue, New York, NY 10017, USA

Routledge is an imprint of the Taylor & Francis Group, an informa business

First issued in paperback 2016

British Library Cataloguing in Publication Data
Arora, Payal.
  Dot com mantra : social computing in the central Himalayas.
  -- (Voices in development management)
  1. Internet--Social aspects--Himalaya Mountains Region.
  2. Internet--Economic aspects--Himalaya Mountains Region.
  3. Right to Internet access--Himalaya Mountains Region.
  4. Computer users--Himalaya Mountains Region. 5. Internet
  and the poor--Himalaya Mountains Region. 6. Education--
  Effect of technological innovations on--Himalaya Mountains
  Region. 7. Rural poor--Education--Technological
  innovations--Himalaya Mountains Region.
  I. Title II. Series
  303.4'834'095496-dc22

Library of Congress Cataloging-in-Publication Data
Arora, Payal.
  Dot com mantra : social computing in the Central Himalayas / by Payal Arora.
    p. cm. -- (Voices in development management)
  Includes bibliographical references and index.
  ISBN 978-1-4094-0107-0 (hbk)
  1. Internet--Social aspects--India--Almora. 2. Computer networks--Social aspects--India--
  Almora. 3. Computer literacy--India--Almora. 4. Almora (India)--Social conditions. I. Title.
  HN690.A455A76 2010
  303.48'33095451--dc22

                                                                                    2010017056

ISBN 978-1-4094-0107-0 (hbk)
ISBN 978-1-138-26059-7 (pbk)

Contents

List of Figures

Table

To
Padraig Seosamh Tobin

Foreword

Mark Warschauer

Professor of Education and Informatics, University of California, Irvine

In a recent discussion of the One Laptop Per Child (OLPC) program's desire to place computers in the hands of every Indian child, one laptop enthusiast was asked how he was so confident that rural children, who may have never been to school, could make productive use of laptops for learning. "Well, after all," he replied, "they are *digital natives*."

Deterministic views of the positive power of technology are seldom as extreme as this; after all, how many people believe that simply being born in the information era will endow certain capabilities on youth, even if they and their friends, relatives, neighbors and teachers, may never have touched a computer? Yet the belief that new technology has some magical power to solve or bypass complex social, educational, economic and political problems is certainly widespread, as witnessed, for example, by the OLPC program itself (for one critique, see Kraemer, Dedrick and Sharma, 2009).

Unfortunately, the most common counter to this naïve techno-optimism is a narrow-sighted techno-pessimism that focuses solely on the problems associated with new media and fails to recognize their potential value in individual and social development. Techno-pessimists deny that those on the margins could benefit from *any* use of computers, since presumably all other more pressing social and economic problems would need to be solved before the poor could productively deploy new technology.

Chris Dede succinctly foiled both the techno-optimists and pessimists with a simple metaphor. As he explained in testimony to the US Congress:

> Information technologies are more like clothes than like fire. Fire is a wonderful technology because, without knowing anything about how it operates, you can get warm just standing close by. People sometimes find computers, televisions and telecommunications frustrating because they expect these devices to radiate knowledge. But all information technologies are more like clothes; to get a benefit, you must make them a part of your personal space, tailored to your needs. New media complement existing approaches to widen our repertoire of communication; properly designed, they do not eliminate choices or force us into high tech, low touch situations. (Dede, 1995)

If, indeed, new technologies are more like clothes than fire, we need to understand how people make them fit their lives, how, in Dede's words, they make them "part of your personal space, tailored to their needs."

Few efforts to do so are more successful than that of this book. Payal Arora takes on a research task that few have sufficiently valued and far fewer have accomplished: becoming one with a community and its people, gaining their trust, examining how they make use of technology according to their own context and needs, and revealing that to the world in all its nuance, biased by neither sentimentality nor judgment. She brings many strengths to that effort, including a multicultural, multilingual background that allowed her to interact as an insider while reflecting as an outsider; a keen observational eye and ear and a willingness to patiently watch, listen and probe; an analytical framework that focused on issues of relevance to both the participants and readers of this ethnography; and a graceful, poetic writing style that makes every paragraph a pleasure to read.

In the end, Arora concludes that the people of Almora are very much like the rest of us in their fundamental attitudes toward new technology. But the path that she takes to get there brilliantly illuminates the *particular* meanings of digital media in the Central Himalayan context, thus making her conclusions all the more insightful and important. In Arora's eyes, and in those of the reader blessed with the privilege of taking in this book, the common humanity that unites the rural poor in Almora with the digitally privileged in San Francisco or London is not based on a denial of our different needs for and uses of technology, but rather on an understanding and appreciation of them.

References

Dede, C. (1995). Testimony to the US Congress, House of Representatives, Joint Hearing on Educational Technology in the 21st century. Retrieved August 17, 2009, from http://www.virtual.gmu.edu/SS_research/cdpapers/congrpdf.htm.

Kraemer, K.L., Dedrick, J. and Sharma, P. (2009). One laptop per child: Vision vs. reality. *Communications of the ACM*, 52(6), 66–73.

Preface

Conducting an ethnography is a humbling experience, reminding us that our lifestyles are but specific choices within specific social worlds of understanding. And reminders are aplenty: they come in the form of entrepreneurial villagers in Almora, setting up *chai* and *butta* (corn) stalls alongside the winding mountain pathways in the freezing cold of the day; passionate NGO workers going village to village drumming up support for their numerous causes from water catchment areas to electric cookers; women and girls carrying firewood on their heads, miles away from home.

More specifically, the Uttarakhand Seva Nidhi Paryavaran Shiksha Sansthan (USNPSS) organization in Almora, India, provided a vital nesting ground for the nurturing of my ideas. Dr Lalit Pande, the director, with his acidic skepticism of new technologies for development, balanced by a relentless honesty and deep dedication to the people of Almora kept me on my toes. Anuradha Pande, with her single-minded dedication to women's empowerment in these hills propelled me to think continuously about the gender implications to events and issues encountered. Champa, Rama, Kamal, GP, Renu, Mohan and others kept me electrified and deeply moved with their range of stories over these years with the villagers – Rama's ongoing "war" with corruption in the local hospitals, Champa's determination to push young girls to ask for more in life, GP with his passion for technology and learning, Kamal with his gentle and inspiring views on life and Renu's love for entertaining with the villagers through songs and theatre, all contributed greatly to this work. I am particularly grateful to Kamal K. Joshi for the generosity of his time and effort in translating select interviews into the Devanagari script. Also, I am thankful to Awadhesh Verma for giving me the rare opportunity to interact and work with high-end Indian policy makers and the private sector regarding rural computerization. My journey to Almora was particularly inspired and arranged by Dr K.S.Valdiya, the famous Indian geologist and native of this region, also lovingly called the "man of the mountains." He serves as a standing reminder that great scholarship is rooted in love for people and place. I am and will continue to be infinitely grateful to all the people I encountered in the villages and towns; students, teachers, principals, businessmen, cybercafé owners, government officials and more who opened themselves and often their homes to me.

Of course, this research builds on my past scholarly immersions over the years that would not have been possible without the help of key institutions and individuals: Dr Brij Kothari, founder of PlanetRead, was an excellent mentor and guide for the hp iCommunity project in Kuppam in 2004. Also, I will always be thankful to Michael Spencer, founder of Hospital Audiences in New York, for giving me the first opportunity to do pure research on a National Institute of Health project in 2002. Further, this book has been paved with ideas that owe its inspiration and support to key institutions and persons: Teachers College, Columbia University has generously supported my three and a half year doctoral journey through financial scholarships, thereby allowing me to focus more on research. The Kellogg Grant provided me with further substantive funding for which I am deeply grateful to Professor Jo Anne Kleifgen for bringing me on board. In fact, Dr Kleifgen has a special place in this process as she has been advising me for a number of years, well before the pursuit of this research. Also, Professor Charles Kinzer has been a pillar of support for which I am deeply grateful. A special thank you is owed to my sponsor, Professor Hervé Varenne, who has been a fundamental influence in my writing and thinking over these years. Anyone remotely familiar with his work can see his deep-seated impact in the kinds of questioning and avenues of critique that I have taken on. His tendency to go against the tide and his discomfort with popular notions and explanations is very infectious, often leading to original thinking in any arena. I count myself greatly fortunate to have gained his continuous mentorship over these years.

Tremendous support from numerous other sources – both scholars and practitioners alike: Professor John Broughton, Professor Arjun Appadurai, Dr Jonathan Donner, Professor Arvind Singhal, Dr Sushil Sharma, Dr Aaron Hung, Dr Karen Kaun, Kati Kabat, Joelle Brink and Jennifer Swaner. In particular, I am honored that Professor Mark Warschauer has taken the time to write a kind and thoughtful Foreword for this book. He was one of the first people to push me to publish when I was just in the masters program and his book, *Technology and Social Inclusion: Rethinking the Digital Divide* was what set my direction of research towards new technology and social practice.

Finally, my family are and always will be my backbone. There is no doubt that my way of expression can be directly attributed to my father, Ranjiv Arora, a natural professor and preacher; I get my inspiration for the love of learning and adventure from my mother, Rachna, and if ever in need for absolute support, my sister, Panchmi, has always been there, through thick and thin. Last but by no means the least, I dedicate this book to my best friend, Padráig Seosamh Tobin, without whom I would be undeniably incomplete.

List of Abbreviations

AIR	Air India Radio
Arya Samaj	Indian political party
Atta	Flour
Balwadis	Pre-schools
Bhagwat Gita	Indian sacred text
Bhoomi project	Technology project in India
Bhotiyas	Tribal people of the Himalayas
BJP	Bharatya Janata Party, main opposition party in India
Brahmins	Priestly caste
BSNL	Bharat Sanchar Nigam Ltd.
Butta	Corn
CBO	Community-Based Organization
Chai	Tea
Chipko	Movement to save the forests in the Himalayas; literally means to "stick"
Chow mien	Chinese food dish
Chula	Stove
CICs	Computer Information Centers
Crore	Hundred million rupees
Damru	Drum
DG	Digital Green
Didi	Sister
DIT	Department of Information Technology
DLIF	Digital Library for Indian Farmers
Doli	Cot
Doordharshan	Government Television Station
E-chaupal	Movement for agricultural markets by ITC
Ghee	Clarified butter for cooking
GIS	Geographical Information System
Harijans	Untouchables
HiWEL	Hole-in-the-Wall Education Ltd.

IAS	Indian Administrative Service
ICT	Information Communication and Technology
Jalagams	Water catchments
Karva chaut	A custom where a woman fasts for the long life of her husband
Kethibadi	Farming
Kisan	Farmer
Lakh	Two thousand dollars
LS	Learning Stations
Mahila	Woman
Mandi	Government market
Mehendi	Henna
MIE	Minimally Invasive Education
Mission 2007	Current national project to create connectivity across Indian rural villages
Neem	Herb
Netas	Politician
NGO	Non Government Organization
NIC	National Informatics Center
NRI	Non Resident Indian
Nula	Water pump
Paan	Beetle nut
Padma Shri	Second highest honor bestowed by the government to its citizens for their contribution to the country
Pahadi	Hill people
Pakodas	Fried Indian snack
Panchayat	Village leaders
Patwaris	Middlemen
Prasad	Holy offering
Rajputs	Warrier caste
Sadhu	Sage
Sangattan	Collective
Sarpanch	Village head
Soochna Kutirs	Government cybercafés
Swami	Sage
Thakur	In this region, this title signifies royal and noble rank

Upanishads	Holy scriptures
Uttara	Government information portal
USNPSS	Uttarakhand Seva Nidhi Paryavaran Shiksha Sansthan

Chapter 1
Introduction

Local as Celebrity

In a village in India, corporate-sponsored students ask the *sarpanch*, the village head, about computers, hoping to gain information on the needs of the villager:

> At one point, one of the lead students asked: "What do you expect the 'computer' to do for your village?" The sarpanch answered: "We are not sure what 'computers' do, so we are not sure what to answer." The student began to explain and suddenly, to our ears anyway, the room erupted into what seemed like all 30+ people – standing, sitting, hanging on in windows and open doorways – speaking simultaneously. At one point, several minutes into this, the other student, sitting near us leaned over and said, "this is getting out of control." Then, just as suddenly the tumult quieted and the sarpanch indicated the room now understood what "computers" could do. They then proceed to provide excellent responses to the original question. (Thomas and Salvador, 2006, p. 112)

The miracle of the "local" has arrived; from supposed passive remote individuals, they are now viewed as receptive social collectives with shared knowledge, craving to chalk their path with new technologies. The local today enjoys celebrity status. In the last decade, corporations, states and transnational agencies alike have unleashed their army of ethnographers to unravel the mysteries of the local consumer, particularly in what is considered remote, disadvantaged and marginalized areas of so called Third-World nations as they interact with computers. India is one of their favorite sites as it serves as a "live laboratory" and potential market for new technology interventions given its Silicon Valley status. The new consumers within this laboratory are now seen as promising subjects that will engage and socialize with new technologies and "leapfrog" barriers of illiteracy, superstition and poverty to access and use these computer-mediated spaces for socio-economic mobility (Bhatnagar and Schware, 2000; Keniston and Kumar, 2004).

Yet behind these anecdotal celebrations, there is little understanding of this supposed leapfrogging: of how people in marginalized areas instruct each other and themselves about and with computers for varied purposes; of how people play with new technologies and navigate and circumvent such artifacts and its spaces to create potentially new forms of products, processes and practices; of people's perceptions and beliefs of computers and its relation with other new and old technologies and its institutions as well the effect of past to current government policies and practices in technology dissemination for development on people's receptivity and usage of computers. Such a focus on day-to-day learning is important if we are to genuinely de-romanticize and demystify constructs of deliverance and promises of computers as pathways to change.

In fact, every new technology gets caught up in the timeless debate of euphoric proclamations of its transformative capacities on social practice to foreboding condemnations of such technologies as oppressive, divisive, homogenizing and regulating of social life. Not surprisingly, in current times we are bombarded with a slew of optimistic discourse on the effects of the computer and the Net on social unison, mobility and democracy: the promise of the "information superhighway" (Cozic, 1996), "virtual communities" (Rheingold, 2000) and "netizens" (Hauben and Hauben, 1997). Simultaneously, there is a legitimate concern of participation within this digital sociality with emphasis on the "digital divide," a perceived phenomenon of technological exclusion based on social, cultural, economic and other parameters (Warschauer, 2003). Hence, there is a need to provide a frame of reference from which to evaluate these various claims made about properties of a new technology, its novelty and its relationship with sociality at large.

Social Learning with Computers

This book investigates *how and for what purposes do people in Almora, Central Himalayas, India, use computers and the Net in informal public venues.* This is an eight month ethnographic work in Almora, a town of 56,000 people in Uttarakhand, Central Himalayas. This study focuses on relations between old and new technologies, how people harness physical, social, economic and cultural resources to facilitate their understandings with these artifacts, the nature and consequences of this learning as well as perceptions and beliefs about the artifacts – its situated spaces and activities within the larger context of development policy and practice. Investigating practices amongst a relatively remote and new group of users of the computer and the Net allows for possible new perspectives to emerge and perhaps old views to be reinforced and revisited. This work contributes to realms of user-interfacing, new media, information consumption and production, social learning with technology, ICT and international development and policy and practice of new technologies for social change.

The book moves beyond the hype of new technologies by delving into the spectrum of human imagination and enactments with computers and the Net. Technology can be viewed as an artifact and technique of human invention that shape and is shaped by social learning with often unpredictable consequences. Technology is seen as a material embodiment of an idea. What constitutes that idea depends on who and where it is framed and to what ends. Social learning is viewed as a dialectic process and enactment of human ingenuity. We cannot make meaning of technology without understanding its place and space, its boundaries, frames of reference, its coordinates of interpretation, functionality and optimization. After all, technology usage is a situated social practice. This study indexes technology in realms of a learning context, however temporarily so, to best understand its enactment.

Figure 1.1   Map of Uttarakhand, India
Source: Rajiv Rawat, www.uttarakhand.net

Almora town is situated in the Almora district of the Kumaon region in Uttarakhand State, Central Himalayas. It is "remote" in a sense of being eight hours away from one of the nearest cities, Delhi, and being relatively isolated in the mountains where households are scattered across hilly terrains making access and usage of goods, people and spaces a significant challenge. With 90 percent of its 632,866 population living in villages and surviving on subsistence agriculture, this area is looked upon as disadvantaged and marginalized and officially demarcated as "backward" (Sati and Sati, 2000). The town itself has around 56,000 people. Furthermore, with the formation of Uttarakhand as a new state in 2000 (carved out of Uttar Pradesh), an influx of national capital has come towards this region to develop it particularly through technology, earning the title of an "e-readiness" state (OECD, 2006). Hence, in the last five years, schools and universities have been provided with computers, with broadband entering this arena only in the last year or so. Besides this, public-private partnerships have blossomed to set up public venues for computer and Net access and usage for "economic and social mobility" of the masses (Garai and Shadrach, 2006). These nascent initiatives contribute to making Almora an interesting site of research.

The deliberate choice to investigate learning practices with computers *outside* school settings is primarily to reveal a wider perspective of education that is not hindered by chronic formal institutional difficulties. A good amount of research has gone into formal educational and technical failures including but not limited to poor connectivity, teacher training, electricity, software and hardware access and design, maintenance, lack of relevant content and more. Rather than continue to circulate such findings by following up with formal education's limitations on learning, this study looks for alternative sites of engagement, in this case at cybercafés, non-governmental organizations (NGOs) and government cyber kiosks. The rationale is that by looking outside school settings in public contexts where engagement does happen, much can be learnt about how people actually interact with these resources in the widest possible sense.

By focusing on the micro-politics of engagement through ethnography of new technologies, much can be revealed about the macro-political nature of technological intervention, mediation and human ingenuity with new tools and practices: how people talk about these new tools, what they say about it, to whom they talk about it to, why they choose to use it at specific moments in time and space, when they seek for it and when they disengage from it, how they relate past and ongoing practices with old and different technologies to these new tools, how they learn to reproduce, modify and transform events with these artifacts, where and how they position themselves and others in relation to these tools, why they choose certain spaces over others, are all factors contributing to the understandings of usage with computers and the Net.

By no means does this study intend to "cover" Almora in terms of representing it as a "local whole;" rather, creates a strategic coalescence of captured experiences and discourses of people using and being used by technology; perceptions to practices:

> Although multi-sited ethnography is an exercise in mapping terrain, its goal is not holistic representation, an ethnographic portrayal of the world system as a totality. Rather, it claims that any ethnography of a cultural formation in the world system is also an ethnography of the system and therefore cannot be understood only in terms of the conventional single-site mise-en-scene of ethnography, assuming indeed it is the cultural formation, produced in several different locales, rather than the conditions of a particular set of subjects that is the object of study. For ethnography, then, there is no global in the global/local contrast so frequently evoked...The global collapses into, and made an integral part of parallel, related local situations rather than something monolithic or external to them. (Marcus, 1998, p. 83)

This pastiche allows for a glimpse of a range of usage regarding computers, through which we can gain insight into pertinent concepts of technological novelty and impact within larger contexts of international development and policy. Less attention is paid to who is seen to be on which side of the digital divide and more on how people, spaces, ideas, activities and tools come together in deliberate and strategic orchestrations for the enablement of the computing event.

Methodology

Eight months of fieldwork yielded a large body of data of which only a small fraction is delved into in-depth for the present discussion. In addition to field notes, an intensive analysis was made of all the literature produced and made available at the local level in the form of government documents and non-government organization (NGO) promotional literature and reports. At the same time, a wide range of material, relevant to the daily activities was amassed: drafts of articles, news abstracts, letters, photographs, emails, memos, and more. Formal interviews were carried out with a large number of people from teachers, principals, students, local government officials, swamis and sadhus, NGO founders and employees, villagers, youth, grassroots activists, local business people to farmers and traders.

Focused group discussions with farmers over week long periods across the duration of fieldwork added a significant dimension to this analysis. All such discussions transpired in Hindi. These interviews supplemented the vast body of comments and information gleaned during informal discussions. Surveys were conducted amongst youth from the urban to rural counterpart reflecting important patterns in perceptions and usage of computers.

Reflections through observation provided a further source of data. No attempt was made to conceal my role as a researcher. For example, when I interviewed people, the tape recorder was clearly present and publicly acknowledged. In group discussions, even when I took a backseat, the group was cognizant of my role as the researcher. In fact, I made it a point to give the participants the opportunity to question me on matters of their interest which did not limit itself to the topic at large. This allowed for mutual curiosity and interest to deepen and shift such interactions from an interview mode to that of a discussion. As with all ethnographies, key names, dates and places have been changed to protect the identity of those involved. In cases where people and institutions have been named, it is to pay my deep debt to them for their significant influence on my work as elaborated in the acknowledgement section.

I chose to be stationed for the duration of my fieldwork at Uttarakhand Seva Nidhi Paryavaran Shiksha Sansthan (USNPSS), a well established and reputed NGO committed to environment and education. This NGO was located in the town of Almora. However, given the past decades of outreach and capacity building, they had succeeded in nurturing ties with the most remote of villages. Using this as a base, I traveled to neighboring villages and towns for a good portion of the time (perhaps a third of the time) based on strategic opportunities and events that came my way. This back and forth between the rural and the urban area allowed for the juxtapositioning of practices and perceptions that was influential in the analysis of this work. Having such a base itself was a tremendous boon. By mere association with this respected NGO, I found people more receptive and open to questions. More importantly, the almost daily discussions with the staff and the founder who had invested a great portion of their lives to the development of this region, added a constant challenge and insight to day-to-day events.

Perhaps the most valuable aspect of my association with this NGO was its weekly training sessions with villagers from Almora and neighboring districts at large. The NGO made a concerted effort to get villagers to their NGO for training which included young girls as *balwadi* or pre-school teachers, farmer cooperatives

Figure 1.2   Researcher (bottom left) with *Balwadi* Teachers at USNPSS

to women groups for local elections, health and livelihood. Through my stay there, I met and lived in the same quarters with these people and thereby was able to have prolonged and deep discussions with them in less formal settings. Of course, there was always the need to be aware of the NGO's own biases when interpreting such data. This tension of learning and yet disassociating from this organization was a constant struggle throughout my time there.

So while the NGO provided a fertile ground to explore perceptions of villagers, the actual identification of computer usage took a rather different path. At the embarking of fieldwork, I was directed to "failing" technology initiatives: the famous Hole-in-the-Wall (HiWEL) experiment by NIIT (an Indian IT company) and the *Soochna Kutirs*, the Computer Information Centers (CICs) for economic and social mobility as part of the national digital equity drive. These two sites of low engagement form Part II of the book under "Computers and Rural Development," followed by "Computing in Cybercafés" in the Part III section, the main sites for high public engagements with computers. To go more in-depth into the workings at cybercafés, I chose one cybercafé reputed to be the most popular. New arenas sometimes require new strategies of investigation. Given its limited space, I volunteered to work for free at the cybercafé during my stay there in exchange for being part of its activity. This action research posed a constant dilemma for me where I became complicit in activities, some of which can be deemed as "plagiaristic." In addition, there was continued effort required to restrain from advising users at this café in their understandings and interpretations of online content.

As in any ethnography, trade-offs are many and dilemmas are aplenty. This study is no exception. The age old tension of breadth versus depth was encountered at the local arena as I chose to pursue multiple sites of engagement and discourse over delving deeply into just one site or issue. By being stationed at an NGO, there is no denying that some of the discussions encountered were colored by its larger ideology which perhaps was projected onto me.

This study offers a partial but important and unique perspective regarding computer usage, helping to challenge some popular notions on computing. It shifts some of the key debates and assumptions on computing: direct interfacing and empowerment, relations between online information and socio-economic mobility, the makings of knowledge and its relations to new technology, the ameliorative role of computers and the role of intermediaries and institutions in technology design, access and usage.

The emphasis on the social reveals my bias over that of a cognitive and quantitative kind. However, that can be said about *all* studies in terms of biases of a varied kind. In this case, we should look at such pursuits as extensions of other such discussions to form a composite understanding of practice. By association and intersection of these investigations, its complexity is revealed to give rise to better and more pertinent questions as well as fairer and more legitimate assumptions of practice. In other words, this work strives to humble large claims in policy and practice.

Techno-Revelations for Development Policy and Practice

This fieldwork with relatively new computer users at cybercafés has revealed that much of their computing activity is arguably non-utilitarian in an economic sense; instead, it is more centered on social and entertainment purposes. This has led to further the investigation in relations between labor, leisure and learning within transcultural and technologically-mediated environments. Also, for the most part, most of these users happen to come from the town versus the villages, although, as will be revealed, engagements come in multiple forms and through a range of critical intermediaries.

The encountering of high-profile national technology initiatives at the ground level has lent an important policy perspective to this study, demonstrating how these initiatives play out. In particular, social interaction with computers has been looked at closely as tied to two such projects: 1) *Soochna Kutirs*, ICT kiosks designed as digital "knowledge centers" for rural people as part of the *Mission 2007* national policy initiative to connect 600,000 villages in India through computers and, 2) the HiWEL (Hole-in-the-Wall) "Learning Stations" for "free" learning for children away from schools. This has led to a deeper exploration on issues of what constitutes as free learning, relevant, global and/or correct information, the range and role of actors that serve as intermediaries of information within specific public spaces/contexts, implications of direct versus indirect access and usage of computers on learning, gender and technology to the consumption and production of knowledge.

To understand computing, it is important to focus on both access and usage of computers. Access is seen to be divided particularly along certain lines, such as rural versus urban, poor versus rich, developing versus developed countries, women versus men and the like. The overused term of the "digital divide" stems from a human rights perspective; the notion that all people, regardless of their contexts and affiliations have the right to access technology so as to be on the same leveling field. Without undermining these territories and divisions, this study strives to highlight the flows *between* these categories and demarcations to avoid reifications that can trap our understandings. Thereby, this work steers away from identity politics as a central means of interpretation of social practice.

That said, this study holds the view that ideas manifest through the design, implementation and spread of technology. How people talk about technology and how they use it rests on the multiple positions/statuses they occupy in society that may or may not play out in totality in the actual enactments with the artifact. While the characteristics of the social actor is acknowledged as a way of understanding how technology is adapted, used, circulated, and played with for a range of outcomes, intentional and unintentional, these positions are attended to only if there is an indication of their influence on action. For instance, far from technology being neutral, its design and purpose can be argued to be deeply "gendered," and yet can simultaneously be perceived through a variety of lens, be it class, caste, and other social factors. That said, if a tool is gendered in a particular way, does it necessarily condition all of its social action? Are there ways of circumventing, navigating and perhaps transforming these gendered aspects into other forms of practice?

Overall, the reader will witness a continuous struggle with the understanding of new technology; its temporal and spatial geographies of occupation, its imaginative possibilities, as well as the ways in which it gains meaning and plays out through social practice. The underlying assumption here is that technology usage is a social phenomenon. Thus questions of causality of technology give way to relational ones: In what ways do people form relationships with specific technologies? What role does the materiality/physicality of technology play in the construction of social practice? What do we mean by affordances and constraints of technology and what are its consequences? If we are to look at technology usage as an orchestration of artifacts, spaces and people, how does that play out in different settings and groups be it socio-cultural, historical to transnational? Such are the concerns under current investigation.

Organization of the Argument

This research focuses on interactivity with technology at a discourse and practice level. What we do not concern ourselves with are questions of whether or not computers are good or bad, but rather the range of constraints and opportunities surrounding computers. The interest here is on how a diverse group of people compute, with the hope that these findings will contribute to larger discussions on new technology in society.

Chapter 2 provides a backdrop of pertinent and dominant literature concerning this topic. It starts with a broad overview of shifts in development policy and practice and relations of technology to social change, leading to the more narrowed lens of technology in India. In other words, by understanding how tools of human endeavor have been perceived and instigated as well as undergone change, we may gain a stronger foothold to situate the findings from this study.

Part I of the book is sectioned to reveal the nature and character of Almora and its people and their relationship with a plethora of technologies, both old and new. Chapter 3 takes the reader on a journey through Almora, recreating some of the flavor and feel of the place, making familiar the topographical and contemporary personality of Almora. The interspersing of episodes and events, of people and places are positioned to give the reader some of the complexity of this place with the key purpose to make messy the dichotomies of the local and global, rural and urban, or for that matter, traditional versus modern ethos. Social beliefs, people's engagements with technology and development initiatives in this area are briefly touched upon here.

Chapter 4 extends the prior chapter by making known how a family of technologies, be it the plough, voting machines, cellphones, ATMs to smart cards, come together in tremendously creative and surprising ways, underlining the persistence, innovations and shifts in social practices with technologies in this area. This lens reminds the reader that the computer is but one tool added to the cauldron of existing technologies, and the unique history of interactivity and engagement in Almora.

Chapter 5 and Chapter 6 come together under Part II of the book – "Computers and Rural Development," demarcated based on its common link with contemporary IT national policy and practice. Chapter 5 closely examines discussions amongst groups of farmers, totaling 160 at an NGO on the issue of "relevant" information as key to socio-economic mobility amongst farmers. This chapter questions the premise of current development policy, emphasizing computers as tools to supplant *patwaris*, middlemen in the agricultural sector, by empowering farmers through direct access to crop prices and other "essential" information. This helps situate the popular *e-chaupal* national technology initiative where farmers supposedly can access such information from the computer. This is meant to help them reach markets directly and sell their produce fairly by circumventing middlemen that serve as intermediaries of trade. Instead, it is argued here that decision-making is not based primarily on "relevant" information. And even if farmers overcome institutional and technical hurdles in accessing computers, these tools will continue to be weak intermediaries in agricultural decision-making until larger systems of agricultural production, consumption and market choices along with equity in access to other concerning institutions and agencies substantively change with the interests of the farmer at the forefront.

In Chapter 6, we move from the agricultural domain to the educational front where a different kind of intermediary is being circumvented – the School. This chapter investigates how the famous "Hole-in-the-Wall" (HiWEL) World Bank funded "minimally invasive educational" model plays out in Almora. Here, computer kiosks are designed specifically with children in mind to provide "free" learning through direct access to these tools away from schools. The idea behind this is that schools tend to restrain children from exploring their true learning potential. Computers are seen to provide a window to learning that is not shackled by the much documented poor state of government schools and their teachers. Instead, this chapter reveals the failure of this project in Almora and investigates the possible reasons for such a downturn. In doing so, the relationship between "formal" and "informal" learning spaces and activities and social learning is explored. What constitutes "free" learning is at the heart of such analysis. The role of formal educational institutions in the democracy of learning gets played out and juxtaposed against HiWEL's ideology.

In Part III of this book "Computing in Cybercafés," Chapter 7, Chapter 8 and Chapter 9 explore the range of activity within cybercafés in Almora town. Having volunteered at a popular cybercafé in exchange to witness these micro-processes at work, I gained insight into the diverse spectrum of activities within such a computer-mediated space. For example, Chapter 7 is concerned with a range of learning that goes into the harnessing of multiple online resources at hand to accomplish schooling tasks at this cybercafé. In doing so, issues of direct versus indirect interfacing with computers, digital learning, plagiarism, collaborative learning and the private-public nature of cybercafés in relation to schools and the larger Indian educational system is explored. It is argued here that sidelining the "immorality" of the "plagiaristic" act allows us to focus on what is more interesting a finding – the ingenuity of students as they go about recognizing, sharing and capitalizing on their environment; people and artifacts at hand to achieve a "successful" learning experience and end product.

Chapter 8 turns to a learning event at the cybercafé where girls shop for Western and Indian painting images at Google for their art portfolio. Their interactions with these images reveal their criteria for what qualifies as "Western" and "Indian" art. This provides a good basis to investigate actual learning engagements with online "global" content. In this process we delve into the nature of information versus knowledge, consumption versus production of knowledge and globalization of content through the Net. The point here is to emphasize that interaction does not necessarily equate with understanding, learning engagements with new technologies can be peripheral and fleeting and that which gets learnt can diverge far from what is expected to be learnt.

Lastly, Chapter 9 reveals that there is an intricate relationship between leisure, labor and learning based on the fact that these cybercafés are primarily sustained by not pragmatic tasks but by what is discovered to be mainly entertainment and social endeavors. From Orkut, a social networking site to music downloads and instant messaging and dating, this space transforms into a recreational hub. The argument thickens as we recognize that much effort is required for leisure as has already been demonstrated in the past two chapters. Furthermore, I conduct a survey across three different intercolleges to investigate the economics of leisure. It is found that youth, regardless of economic status, share common ground on career aspirations as well as the will and capability to play with new technology. Overall, utilitarian notions of development are challenged, gently reminding us that poor people do have a lot more in common with the rest of us, more so than we credit them with. Leisure is a demand and a necessity amongst one and all.

Chapter 2
Frogs to Princes: Taking the Leap

The Pathway to Good Intentions: The Development Story

Since the 1990s, social and economic development has carried notions of participation, empowerment and sustainability. Here, the "locals" are regarded as critical resources to their own development instead as subjects to be studied; as such, sustainability in development is believed to be achievable only when people are deeply engaged in facilitating change within their communities (Baker, 1997; Blewitt, 2008; Putnam, 2002). In fact, it is fashionable to subscribe to the Freirean ways of human communication and dialogue, where people conjointly embrace "critical consciousness" to enable true change in their day-to-day lives (1986). While commonly accepted today at least at the rhetorical level by policy-makers and practitioners, these notions emerge from decades of struggling to move away from the victimhood discourse.

Facing chronic poverty, most post-colonial nations adopted well-meaning schemes devised to help them *catch-up* with the so-called developed nations. In essence, developed nations were meant to serve as role models for developing countries. After all, the argument went, there is an evolution of society from "primitive" to "modern," that necessitates the abandoning of old thinking, of culturally backward views and practices and the embracing of an enlightened higher order thinking based on scientific rationalism (Peet and Hartwick, 1999). This became synonymous with westernization and modernization, two powerful concepts that was reified through policy-making across decades. This manifested an "us" versus "them" paradigm; of First versus Third-World people; of dichotomies including rural versus urban, traditional versus modern, inward versus outward looking, and cultural versus economic modes of development. The "development project" was marked by a new universalism, with such values deemed as essential and normative to all concerned:

> The linking of human development to national economic growth was a key historical event. This is why the term development project is useful. It was a political and intellectual response to the state of the world at the historical moment of decolonization. Under these conditions, development assumed a specific meaning. It imposed an economic understanding on social life. This meant that development was a process that could be universal and should be unimpeded by specific cultural patterns. Its two universal ingredients were the nation-state and economic change. (McMichael, 2004, p. 32)

With "rationality" being the byword for progress, pillars of science and technology were seen as key supporting structures of such momentum. In gauging the betterment of societies, quantifiable measures were readily embraced and propagated, giving birth to national and global indices and statistical methods of growth. This served as the basis upon which decisions with global implications could be made. Local practices were seen as a hindrance towards the scaling of development schemes for national progress.

However, every ideology gets humbled by reality. Deep divides along lines of education, healthcare, unemployment and other agreed upon standards of progress continued to persist decades later. Many saw this as a result of dominant paternalism and negation of local dynamics, and capabilities. Another popular argument is that the overemphasis on the economic realm as a solution to poverty circumvents larger issues of historical colonial exploitation (Frank, Chew, and Denemark, 1996; Goldthorpe, 1996; Sen, 1999). The systematic dismantling of working institutions and agencies of nations during colonial rule and the continued dependency post-independence is seen to have placed these post-colonial nations at a different starting point from that of the "First" world. Thereby, the "metropolis-satellite" or "core-periphery" relations of power are viewed as permeating all sectors of social life. In taking responsibility for the past, development becomes less of an altruistic act on the part of developed nations. Instead, it is portrayed as the right of poorer nations to rebuild their institutions to genuinely compete at a leveling field.

This critique shifts emphasis from the outlook of deficiencies and cultural and social "backwardness" of developing nations to the affirmation of developed nations to make amends. Postmodern arguments take this further by challenging the very notion of the "development project." This battle is fought on the grounds of discourse – definitions, constructions and categorizations. In doing so, we see these divides between the poor and the rich, the rural and the urban, between gender, caste, class, literates, and more as manufactured and sustained by the ruling sector of society to legitimize their control (Bhabha, 1990; Escobar, 1995; Foucault, 1977). As such, these essentialisms become truth to which we are held at gunpoint. It is important to note that this tremendous force and backlash in theorizations was marked for the first time by primarily an intellectual South Diaspora and Third-World authorship.

While this scholarship does not succeed in dismantling the development field, it definitely contributes significantly to the broadening of development parameters to encompass social and cultural factors and not just the economics of nations and their people. In taking into consideration the socio-cultural conditions of the local environment or context, the acceptance for more than one way of development has become widespread (Cheater, 1999; Drèze and Sen, 2002). Participation and empowerment is now the normative vocabulary of contemporary development practice.

So what dictates current policy-making? Today, the dominant neoliberal view in policy entails a more decentralized approach to development while continuing to subscribe to the common goal of helping Third-World countries gain First-World prosperity (Harvey, 2005).What is interesting here is the hybrid stance of tailoring practices and embracing local dynamism while at the same time re-centering economic growth in the development paradigm. Neoliberalism seeks to move beyond what it considers arm chair pontifications of the colonial and postmodern kind by embracing a more action-oriented strategy for global prosperity. Its appeal to policy-makers lies in its potential to artfully amalgamate the local and global and the cultural with the economic into a workable plan of action. Its optimism is best captured below:

> The key point for these countries is that there are practical solutions to almost all their problems. Bad policies of the past can be corrected. The colonial era is truly finished. Even the geographical obstacles can be overcome with new technologies, such as those that control malaria or allow for large crop yields in marginal production areas. But as there is no single explanation for why certain parts of the world remain poor, there is also no single remedy. As I shall stress repeatedly in the pages ahead, a good plan of action starts with a good differential diagnosis of the specific factors that have shaped the economic conditions of the nation. (Sachs, 2005a, p. 50)

In fact, an exemplification of neoliberal policy lies in the much marketed Millennium Development Goals (MDGs), the prime rubric for current global development. It promises an end to poverty by expanding aid for basic human rights: access to quality education, healthcare, appropriate information and communication technologies (ICTs), livelihood, nutrition and political representation. It adopts a multidimensional view of poverty (Narayan, 2000), resting on the now sacred approach of participation. What's more, participation when seen in relation to markets, takes on the role of consumption (Prahalad, 2005). The model of the poor as less of a "beneficiary" and more as a "consumer" voting with their wallet has captured the imagination of many policy-makers and practitioners alike. It promises the beginnings of reconciliation between seemingly long opposing values of profit-making and social good; of the old dilemma of market versus State, of capitalism versus welfare. The byline is that development can be a win-win for all.

Furthermore, participation here is not just with locals but involves the expansion of the playing field to numerous actors, including NGOs, the corporate sector, transnational, international and national foundations, the State and community actors. Public-private partnerships has become the norm as development projects continue to grow in ambition and scope (Osborne, 2000). But more doesn't necessarily mean better. This multilateral participation comes at a price. A noble idea can become pure rhetoric under pressures of day-to-day policy implementation (Cleaver, 1999). Formalism and routinization is often the result of any idea becoming institutionalized. Making the villager the "expert" can, at times, unintentionally transfer a disproportionate burden onto the beneficiary who already shoulders responsibilities for survival (Fisher, 1997). More importantly, this can glorify the self-sufficiencies of the local, entrapping them into the status quo and perhaps even lend credibility to harmful local practices (Wignaraja and Sirivardana, 2004). It can produce alienation and fragmentation amongst a community, as no community should be viewed as a monolithic and homogenous unit. After all, there is always a trade-off. Caste may clash with class or perhaps gender issues may arise at the price of economic prosperity. The harmony of development is but a myth. Development is, after all, unpredictable in its outcome.

Moreover, the recent enthusiasm of the NGO as a more trustworthy intermediary for the poor comes with limitations. These organizations representing the poor needs to be seen in the larger context of their ties to funding agencies and their unique culture and drive, independent of the populations they serve (Barrow and Jennings, 2001). Hence, an NGO can be an extension of the State, a guise of corporate marketing or an actor enacting the interests of a transnational agency more than the needs of its clientele. Similarly, much can be said about the State, corporations and large international organizations that often determine the flow of money and people and thereby, agendas. In fact, current policies are often less a matter of best practices converging as consensus and more on archaic and dominant institutions driving these agendas by their organizational interests and internal politics (Samoff, 1974, 1995).

That said, participation is here to stay. The voices of the poor do not always get heard. A concerted effort is needed to *make* them heard. However messy this process gets, it's still viewed as a better strategy to social development than top-down decision-making. The challenge, however, is to move beyond lip-service to more concrete and localized understandings of social practice. Special attention needs to be paid to intermediaries, technical and human, formal and informal, that are expected to facilitate participation for social good. Further, old ideologies do not disappear but more likely find their way into new discourses and practices, mutating to a more contemporary form. This tells us that some ideas confound deeply. While it is tempting to dwell on the colonial past to explain policy, there is a danger of self-victimization, negating the dynamism of cultures and peoples. Thereby, this work chooses to stay in the present for the most part and, when it goes down the historical trajectory, it is to underline relations with old practices and forms of social organization. In short, as long as actors work towards doing "good," they will continue to be skeptically watched. Yet, the show must go on.

New Technology and Social Change

There is good reason why we put our faith in computers. Like all new technology, it comes with a new promise. It speaks of change. It references the future, not the past. It is neatly packaged as being autonomous, vying for central position through the gravity of its artifact. We live through technology and often live by it. So it should not surprise the reader to see the celebration and glorification of computers and the Net translate itself as a utopia for community, for belongingness, for a new kind of citizenship and world order in the making (Castells, 2000; Drucker, 1995; Kenway, 2006; Rheingold, 2003). Technology is seen to mediate human behavior and bring the world closer through flows of ideas and ideologies; products, processes, people, economies and places. And with every new utopia comes a new ruling class. But for a change, the power is seen to be in the hands of the youth, the "digital natives" of this new and appealing land (Prensky, 2001). It is argued that what constitutes as natural to the youth as they go about engaging with multiple technologies in their digital habitat are often considered alien and threatening to the adult. But instead of fearing youth, a case is made to harness their potential through the best of collaborative behavior. With this, the future is promised to be secure in the hands of these digital aficionados.

And with utopias, comes the altruistic act. After all, if you have something good, some say (with the best of intentions), why not spread it around. Religion, education and democracy are but instances of historical altruisms interpreted and enacted across cultures and continents, with some effects more desirable than others. Besides, the Good Samaritan does not look away; he engages. Hence, the frothy debate of the "digital divide" alongside other well-meaning debates on equity in education, healthcare and livelihood transpire. In essence, social divisions are seen to create and extend technological divisions. The battle falls alongside boundaries; of defining equity along lines of access and usage, hardware and software, language, content and intent that biases the makings of technology and its possibilities.

Regardless of how you slice the cake, many see the access to new technology, in this case computers, as necessary and a basic human right for one and all. This has triggered a flow of capital to developing countries, to help them "leapfrog" barriers of caste, class, gender, poverty, literacy, language, and social and cultural marginalization. This is seen as a means to crack the code to the understanding of emerging markets and the mitigation, if not eradication of the "digital divide" in narrow terms and global poverty on a more grand scale. Thereby, those populations who do not have access to new technology are being heavily invested in, particularly the "next ten percent" (Thomas and Salvador, 2006), "the bottom of the pyramid" (Prahalad, 2005) to the new "netizens" through the "one laptop per child" initiative, a child-centric approach of affordable laptops in poor nations for direct access to knowledge (Negroponte, 1995). Behind this momentum, however, is the presumption that economic benefits do accrue through the access and usage of computers. Also assumed is that locals are "users" or potential users of technology, a naïve but demanding public for computers for utilitarian purposes (Oudshoorn and Pinch, 2003). However, by taking this instrumental view, much of the local or apparent non-users are excluded, giving a narrow perspective of who and what constitutes as local social practice with new technologies.

That said, few would argue against the telegraph, telephone, the printing press; the electric bulb, vaccinations and numerous other ingenious inventions of human effort across time and space that has shaped our society forever. Some go as far as to explain the fate of societies through the timely and strategic embracing and pioneering of new tools and processes, leaving the others behind in the social evolutionary race (Diamond, 2005, 2006). Others are deterministic too in their outlook but far less optimistic. In fact, Marshall McLuhan, the media guru from the 1960s, shaped the doomsday scenario of new technologies into entire disciplines and controversies. He proffered a new vocabulary and conceptualization of communication, ambitious in its scale and sweeping in its generalization of percepts of society impacted by new technologies.

He proposed that new technology affects cognition through the extension of the senses, which in turn affects the consciousness of society as a whole. Much in line with modernization theory, McLuhan described an evolutionary model of technology's influences on society, where the advent of print technology, for example, contributed to the makings of the modern and Western world. Techniques like the movable type credited to the alphabet system is seen to have fostered the chief qualities of individualism, democracy, Protestantism, capitalism and nationalism, key to this new modern "mentality" (McLuhan, 1962). Far from being a technophile, McLuhan meant to warn against such technological changes that configured society into a "global village," where new technologies could tyrannize the social psyche:

> Instead of tending towards a vast Alexandrian library the world has become a computer, an electronic brain, exactly as an infantile piece of science fiction. And as our senses have gone outside us, Big Brother goes inside. So, unless aware of this dynamic, we shall at once move into a phase of panic terrors, exactly befitting a small world of tribal drums, total interdependence, and superimposed co-existence. [...] Terror is the normal state of any oral society, for in it everything affects everything all the time. [...] In our long striving to recover for the Western world a unity of sensibility and of thought and feeling we have no more been prepared to accept the tribal consequences of such unity than we were ready for the fragmentation of the human psyche by print culture. (p. 32)

As the rhetoric of globalization marks the 21st century, so does the emphasis on information and communication technologies (ICT) as a means to accelerate, converge and transform the contemporary human condition (Held and McGrew, 2007). This has provided impetus for scholars to argue over its impact: from *hyperglobalizers* who believe in a new human epoch without borders, shaped by technology for inherently better socio-cultural and economic outcomes (e.g. Sachs, 2005b); *skeptics*, on the other hand, use historical evidence to demonstrate the unremarkability of this so-called new phenomena of globalization, underlining the life altering impact of technologies as mythical. Instead, they emphasize the persistence of the human condition (e.g. Hirst and Thompson, 1996); last but not least, the *transformationalists* who recognize that the deterritorializations of this new era are accelerated by new technologies, but, instead of believing in the converging of societies into a monolithic whole, they believe that such prefiguring and re-engineering of technical interventions are dynamic and open-ended (e.g. Appadurai, 1996; Giddens, 2000; Sassen, 2006a):

> ...inscribed with contradictions and which is significantly shaped by conjunctural factors...of global stratification in which some states, societies, and communities are becoming increasingly enmeshed in the global order while others are becoming increasingly marginalized. (Held, 1999, p. 8)

The fact remains that regardless of one's personal disposition to the impact of new technology, one would be rather naïve to believe that its intervention doesn't come at a price. Hence, tools of war are also used for instituting peace; medical cures can also create new maladies and formidable new problems; the Net can be a space for organizing good as well as a source and confluence of new and renewed hatreds; the mobile can enable medical diagnosis and treatment in areas of least access to healthcare and facilitate political upheavals while, at the same time, be a new means for organizing global terrorism; new technology can provide new lifestyles to age-old communities at the price of losing its historical practices and indigenous knowledge. The communal aspect is countered by its dehumanizing character. Thereby, its potential and intent should not be confused with its usage. Consequences of new technology are limited only by the human imagination.

It is important to note that technology is not just a tool or technique but a process of innovation, invention and interaction. Take for example the famed argument of the "technology of the intellect" – literacy, marking the shift from oral to a literate mode (Goody, 1977):

> In this way a wider range of "thought" was made available to the reading public. No longer did the problem of memory storage dominate man's intellectual life; the human mind was freed to study static "text" (rather than be limited by participation in the dynamic "utterance"), a process that enabled man to stand back from his creation and examine it in a more abstract, generalized, and "rational" way. (p. 37)

This perspective comes with a unique and compelling history, well worth keeping in mind when examining other technical modes and mediums of human construction. For instance, if we are to start from the modernization era of the 1950s and 1960s, the dominant thinking generated during this time was that literacy has certain positive cognitive, political, social, and/or economic "effects" that would transform society and its individuals through what was deemed the most egalitarian means possible – the secular written mode (Graff, 1979). In this line of thinking, literacy is seen to serve societies in transition from their "primitive," "traditional," "backward" states to the "modern." This encompassed ideals of individualism, rationality, linearity of thought and behavior, and knowledge accumulation and objectivity – a shift from myth to science. This view signifies that the transition from an oral to a literate society is the norm. While this theory of literacy has been tempered over these years, the ahistorical, decontextual and consequential shifting of modes continues to be deliberated seriously even today (see Goody, 2000; Olson and Torrance, 1991; Ong, 1988). Further, the privileging of new modes over the old has not lost its proponents. In fact, it has formed a stronger force with the advent of the "information era" of the 21st century (Gilster, 1997; Tyner, 1998). In other words, the belief in the evolutionary nature of literacy from oral to print has new impetus through the digital mode.

However, this view serves as fodder for "new literacy" or sociocultural scholars to critique (Barton, Hamilton, and Ivanic, 2000; Gee, 1996; Wagner, 1999). Scholars, particularly from the anthropology bent, argue that the oral mode has never really been replaced by the literate mode. Instead, what is seen as more useful is to examine these modes as they interact with each other to better understand their relationship in the shaping of society. Further, they debunk the idea that you can talk of society as a monolithic unit with a "collective" consciousness; after all, cognition is the property of an individual and not that of an entire culture. Most importantly, this sociocultural resistance is channeled towards the notion of technological determinism, criticizing that it leaves little room for human agency and cultural context. As such, these "multiple literacies" or social practices are very much contingent on context and power relations rather than a set of universal skills which may be applied to all contexts. Thereby, these scholars debunk this neutrality of literacy, demonstrating through a rich and detailed body of ethnographic studies how, "socially constructed technologies are used within particular institutional frameworks for specific social purposes" (Street, 1993, p. 9).

One can see the shift in thinking in this "great divide" debate through the nature of questions that scholars have posited: from inquiring on the consequences and impact of technology on entire societies' cognitive, social and developmental frameworks to questions situated in the way these mediums interact with actors in specific contexts. Hence, an alternative view of human action is suggested, where actors "play" with boundaries, navigating, appropriating, teasing, contesting and conflicting these arbitrary "artifacts of history" and "products of culture" (Varenne and McDermott, 1998, p. 151). This is complimentary to the social learning and constructivist theorists such as Lave, Latour and others who deviate discussions about properties of individuals, modes, artifacts, to that of "movements" and "networks" between these entities.

In essence, some scholars celebrate the advent of "new" technologies as transformative and revolutionary, much oriented towards technological determinism. Others focus on individuals surrendering to technology, conditioned by those in power and subject to the elite modes of perception and reproduction. Meanwhile, social constructivist scholarship treats individuals as interpretive and innovative beings, able to hybridize and homogenize, appropriate and transform "imposed" technologies into stories of their own making. Here, attention to constraints is just as important as opportunities for change. The argument is made that the novelty of the "new" in technology has been greatly exaggerated and dismissive of histories and cultures. To engage in a symmetrical way of talking about the pre-modern and modern, the old and the new, social constructivists do not deny the differences between the two but rather attribute these differences to scale instead of principle (Latour, 1993). In fact, the "old" can and does exist with the "new" in deliberate contradiction that can be harmonious as well as contesting.

To summarize, as new technologies come to the fore, it is tempting to get swept by the euphoria of novelty, looking for answers within this innovation to ancient and chronic concerns of human communication and human learning. The doctrine that new technologies have unique and often meliorating characteristics and consequences on human life, determining cultural values and mindsets, is seen to lie at one end of the debate spectrum. The other end rejects pegging of human communicative activity along lines of competencies and task-oriented accomplishments. Instead, there is an embracing of complexity, richness and historicity of human life that is socially situated and contextually embedded, defying linearity of change. In other words, we have on our hands the battle between technological conditioning versus social construction with technology, with this book biased towards the latter.

Anthropology of the Artifact: Contexts – Communities – Conducts

Today consumerism is seen as the new participation. The corporation, the transnational organization and the State, through private-public partnerships, are striving to cater to the local by sensitizing themselves to persistent local traditions. In the last five years, IT corporate initiatives have made efforts to transcend disciplines – engineers, designers, sociologists and anthropologists sit at the table to figure out users' "culture" and its effect on new technology adoption, adaptation and design. One can argue that the golden era for the ethnographer has finally arrived. Never before has an ethnographer been placed so clearly at the center of attention in the field of information technology to show the way. The anthropologizing of such phenomena, of recognizing constraints within seemingly borderless and relative spheres of activity are now slowly but steadily reclaiming its space in this discourse. Old habits die hard though. Both private and government institutions are still subject to the normative vocabulary, interests and practices of their respective fields from focus on commercial design for scalability and profit in "user-interfacing" in the private sector, to interventions for "empowerment" in line with popular policy rhetoric in the public sector.

That said, it is important to reiterate that the local is a plurality, potentially conflicting and possibly temporal in its interest and inertia. Further, the local is not necessarily the opposite of global. Parochialism can be found just as much in a global space. In fact, we can argue that context is in the eyes of the beholder. Even so, it is crowned and re-crowned as king to the understanding of social practice. Ever since scholars shifted from the study of actors to that of actions to understand communication, context has become of central preoccupation with often much dispute. It's all about boundaries. As lines are drawn, territories are mapped, and with such institutionalizing of cartographies, seeds for further resistance emerge. The new rhetoric of globalization is just another excuse for re-examination of such boundaries.

This seemingly innocent intent comes from a history of struggle in capturing notions of context. After all, what stays in and what remains out are deeply political acts. Through this lens, context is viewed as that which not just penetrates but "determines the very structure of the interaction" (Goffman, 1966, p. 64). From cognitive to the social, context can be seen as interactional, multidimensional and hierarchical (Duranti and Goodwin, 1992). In a sense, it is not what's in the head but on the ground that matters here, that which is between and amongst people. Preoccupation changes: unfolding of the human mind is overtaken by the capturing of human efforts, enactments, and engagements. By moving from the world of verbal to that of the social, dialogue becomes discourse. Bourdieu's (1984) "field" of social space, where actors or agents position themselves, is undeniably a valuable contribution to our conceptualizing of context but for the implied dispositions and ingrained societal norms that misleadingly claims to determine social action. But let us not negate such valuable efforts that succeed in circumventing chronic and claustrophobic dichotomies of the global-local, public-private, rural-urban, online-offline, and primitive-modern alignments. Taking this further, context can be reframed as that which is situational, institutional and ideological. In other words, context encompasses the micro-politics of an event while at the same time, macro-political in its aspect of stability. In this book, however, the interest is less on the "universal" and more on the "ritual" of practice (Goffman, 1967), less on "domain" and more on "code" (Labov, 1980), less on "text" and more on "context" (Bakhtin and Holquist, 1981). That said, an important question begs answering – if practice travels, does context move with it?

From the physical location of an event to the abstract situatedness of an idea, Anderson's (2006) famous "communities" cross local and national boundaries through the sheer force of imagination yet problematically with little friction. This momentum is later recaptured through the much referenced scapes and flows of ideas, people and things; of global portability with local packaging (Appadurai, 1996). This serves a significant shift from the discussion of context as locational to that of movement. Yet, we are thoughtfully reminded that we can't easily dismiss physicality altogether. Sassen's "global cities" alerts us to the strategic transactions amongst primarily the upper echelons of society in areas of media, finance and other social "cosmopolitanisms" (Sassen, 2006b). So, while some privileged places participate in these porous worlds, others remain unpenetrated.

In fact, our classic romance with "place" appears to give way to the new love affair with "space," accelerated by the onset of new technologies. However, there is a need to rein in these free flowing forms of communication. The problem with contemporary understandings of context lies in its expansiveness and temporality to render it, at times, meaningless. The lack of anchoring as we grapple relations of globalization with new technologies, promising wider forums of practice is disorienting, leaving the local, at best, as a confounding variable. Hence, it is useful to architect social interaction between the artifact and the person through the anthropologizing of context. In doing so, it is tempting to reroute our thinking through the "glocalizing" of context (Featherstone, Lash, and Robertson, 1995), to "think global and act local" as a possible way out of floating theorizations. After all, acting globally can be seen as an impossibility, as the social act is endemically localized. So, what is up for debate is the nature of intent, interaction and consequentiality.

Thereby, ethnography allows for a deeper exploration of human activity, connecting it to broader social ecology within which cultures, cyber and otherwise contest, circulate and cooperate. Escobar puts forth some compelling ethnographic questions that current anthropologists should be concerned with when exploring new technology:

> How do people relate to their technoworlds? If people are differently placed in technospaces (according to race, gender, class and geographical location, "physical ability"), how are their experiences of these spaces different? What are the discourses and practices that are generated around computers? Will notions of community, fieldwork, the body, nature, vision, the subject, identity, and writing be transformed by the new technologies? What continuities do the new technologies exhibit in relation to the modern order? What kinds of appropriations, resistances, or innovations in relation to modern technologies (for instance, by minority cultures) are taking place which might represent different approaches to and understanding of technology? What happens to non-Western perspectives as the new technologies extend their reach? More specifically, which modern practices – in the domains of life, labor and language – shape the current understanding, design and modes of relating to technology? (Escobar et al., 1994, p. 214)

Also, Hess (1995) makes the case that the effects of new technology on Third-World groups are insufficiently understood. Basically, studying technology in anthropology is not new; what is new is that technology is no longer seen as a tool but as a social construction process, "from decontextualized teleology to technologies as complex technosocial systems" (p. 123). Further, it is believed that the production of subjectivities that accompany the study of new technology can only be assessed ethnographically (Turkle, 1995). More specifically, the study of how people relate actively to new technology is seen as being revealed most effectively through the ethnographic process. Also, vernacular usages, movements and cultural politics of such artifacts across the dimension of social power has also gained importance (Gupta and Ferguson, 1997). Thereby, ethnography examines the admixture of people, artifacts and techniques that make up the technosocial event.

In moving beyond Geertz's (1973) "thick descriptions," which reminds us that we study *in* villages and not villages, and extending it to the study of "villageness" in the Bakhtinian sense, of Marcus's multi-sited ethnographies, we trade details for relations. However, if the intent is to capture the politics of movements as not that of typicality but of possibility, there is a chance of representing complexity more faithfully. Either way, the notion of "thickness" of an ethnographic event expressed only through the micro-political, needs to be questioned. After all, no research methodology is authentic in portraying a complete reality. In this study therefore, the event is the sum of social interaction encountered as well as representations of movements, time, space and text. The hope is for a glimpse of multiple realities or "alternative modernities" (Gaonkar, 2001) that serve to understand contradiction, complexity and creativity of new technology practices in a remote postcolonial context, in this case, Almora, India.

Moving through context and community, we come now to conduct. In seeking to understand the interaction of people and computers in Almora, India, it is useful to enter the realm of social learning, the everyday movements embedded in a broader sociocultural context (Lave and Wenger, 1991). Within such asymmetry of spaces, there is a range of relationships and enactments that needs to be brought forth. In centering this learning in "legitimate peripheral participation" of the locals, it:

> ...provides a way to speak about the relations between newcomers and old-timers, and about activities, identities, artifacts, and communities of knowledge and practice. It concerns the process by which newcomers become part of a community of practice. A person's intentions to learn are engaged and the meaning of learning is configured through the process of becoming a full participant in a sociocultural practice. This social process includes, indeed it subsumes, the learning of knowledgeable skills. (p. 29)

To summarize, there is a need to embrace the mapping of movements, relations, and discursive practices to sense the boundedness of people, things and places. Local is neither autonomous nor discrete, but that which is situated within interconnected spaces and topographies of power. This study brings together the spectrum of human innovation in relation with "new" technology: *materially*, in terms of how "new" and "old" technologies come together; *symbolically*, representations and constructions of technology; *socio-culturally*, unique and diverse meaning-making with such artifacts; and *historically*, where personal, public and institutional histories reveal the extent of "novelty" of social praxis with new technology. Of course we cannot assume that the local will employ all of this knowledge. Just because new technology exists within such spaces, it doesn't mean that people will engage with it. There is a need to move beyond the "user" paradigm that dominates the current field and aim for a more inclusive approach to social practice with computers. This study aims to create a meaningful pastiche of social learning with new technology, making concrete the translocal, institutional and global discourses on new technology. This work bridges the micro-political with the macro-political through ethnography of computers in Almora.

Human Ingenuity, Technology and Development in India

> His workshop floor is a swamp of cardboard strips hacked from salvaged boxes. Laborers scoop them up, work them over and give them new life as smaller boxes, which Khan then sells to stationery and packing companies. In another warehouse a few doors down, dozens of rubber soles cut from discarded shoes also await a second chance. Next to these, a mountain of plastic castoffs – toys, computer keyboards, car parts – is separated by squatting workers, to be melted down into tiny pellets before being reborn in some new form. One man's junk is another's fortune. (Chu, 2008, *Los Angeles Times*)

Gandhian self-sufficiency and human creativity finds its image in the Indian entrepreneur. Amidst the twisted lanes of human design in India's slums, to village towns, lies a tremendous amount of ingenuity marked by chronic human need. Entrepreneurship is not bound so much by social good but survival. Desperation and innovation converge to a winning combination. This is manifested through the umpteen types of *bazaars* – markets of goods, people and information, big and small that mark the Indian economy. From slum merchants, rural farmers, to computer geeks in Bangalore's Silicon Valley, the Indian entrepreneur comes in a range of flavors. They are seen to perceive opportunity and act on it in a myriad of creative ways. Opportunity begets opportunity. They are celebrated as the backbone of the Indian society, performing their day-to-day miracles with or in spite of government institutions and agencies.

Entrepreneurship is hardly new terrain. Battling the "License Raj" of bureaucracy, an inheritance from India's colonial past with the British, is a human accomplishment. High trade barriers, taxes and draconian regulation of private enterprises were emblematic of post-independence days, argued to be mitigated by liberalization reforms of the late 1980s (Panagariya, 2008). Under the leadership of Rajiv Gandhi, the then Prime Minister and Manmohan Singh, his financial minister (and current Prime Minister of India, since 2004), the stranglehold on entrepreneurship is said to have been finally eased. This struggle was no small feat. The "double hydra" of nation building (Das, 2000) comprised of Gandhian self-reliant villages with Nehru's more urban and modern visions of industry and science. This led to a dangerous compromise – the License Raj:

> Gandhi distrusted technology but not businessmen. Nehru distrusted businessmen but not technology. Instead of sorting out the contradictions, India mixed the two up and created holy cows. (Das, 2000, p. 2)

India's "holy cow" – modern socialism, while muddled in its views on industry and politics, was very clear on the role of technology to shape its citizens. Nehru's vision won hands down. Citizenship and nationhood entered into a long-standing marriage, bound by the hands of science and technology. Predictably, State education became the grounds for such socialization. While the cultivation of a "scientific temper" has been prized, Gandhian principles have been rewritten to weave it with "modern" values (Arora, 2009). Computers and mobiles for self-sufficiency in this information age, one can argue, has allowed India to rework the Nehru-Gandhian duplex for contemporary times. Thereby, it should be no surprise that digitalization of the nation rests on the village "entrepreneurship" model of scaling computers across India.

Participation has found its friend in enduring Indian entrepreneurism. Gandhian localism and Nehruvian technocratic Statehood comes together once again. The *Mission 2007* policy launched in 2004 commits to connect India's 600,000 villages through 100,000 computer kiosks. These kiosks are popularly called "Community Information Centers" (CICs), manned by village entrepreneurs. The intent is to open up the realm of possibilities for e-government, e-literacy, telemedicine, e-commerce and other information services by leveraging on advances in ICT, particularly in remote towns and villages in India. This massive undertaking is founded on a public-private partnership of the State, technology corporations, transnational agencies and foundations, NGOs, and local governing bodies or *panchayats* coming together.

This basically entails setting up government cybercafés at the confluence of villages or local towns. Village entrepreneurs are expected to sustain these centers by sensitizing to the needs of the local community. Simultaneously, multiple experiments are being generated at the local to State level in areas of technological capacity and connectivity, content generation and knowledge dissemination, resulting in projects such as the digitalization of land records, healthcare and education to pensions. The objective in setting up such centers is "to empower rural communities, create equal opportunity, foster income/employment generation and in general, human development through high economic and social returns" (DIT, 2005, p. 16). In other words, computers are meant to facilitate the much hoped for "leapfrogging" of chronic socio-economic barriers in true entrepreneurship fashion.

However, such a jump has to be by leaps and bounds to truly overcome the multiplicity of hurdles. After all, with more than 50 years of independence, two-thirds of India's population continues to reside in villages while earning less than one-fourth of the national income. These barriers are daunting. Whilst the average growth rate of urban India is about 20 percent, the rural countryside is a mere 2–3 percent per year. India's village arena suffers from a plethora of issues, including high mortality rates, poor nutrition, low access to clean water, poor quality education, corruption, unemployment and overall, appallingly low socio-economic mobility (UNDP, 2009). Almost every field has abysmal ratings: in education, 25 percent of the teachers skip work, and when they do show up, half do not teach at all. Since most of the educational budget goes to teachers, this kind of absenteeism is a tremendous barrier in creating educated youth. In healthcare too, there appears to be little accountability with 35 percent of doctors and nurses not showing up at local hospitals. Also, India suffers from a massive "babuism," a large and draining civil force of paper pushers that keeps the License Raj legacy still burning (*Economist*, 2008).

In this climate, it is natural to be distrustful of government-initiated technological schemes. For instance, it is hard to claim amnesia to systematic State coercion through sterilization of many of India's poor in the 1960s (Hartmann, 1995). This unjustifiable force against the vast poor under the then leadership of Indira Gandhi came from the good intent of controlling the population for India's sustainability and prosperity. Instead, this led to an aversion to family planning for decades and became the "Achilles heel" of Indian politics. Hence, an expensive lesson is learnt: the embracing and diffusion of new technologies comes at the acceptance and *willing* adoption of users.

Tracing our steps further into India's past, we see similar behaviors and consequences. For instance, the usage of the radio by some well-meaning British in the early 1930s for social awareness and learning was, for the most part, perceived as condescension and paternalism (Zivin, 1998):

> The new mass medium – the "great boon of modern science" as one enthusiast put it – was to be employed to keep the Indian peasant content in his natural habitat. Instead of families gathering hearthside around radios in the privacy of their own homes – the picture of bourgeois solidity so enthusiastically promoted in Britain – Indian listeners would congregate in the village square or headman's courtyard to hear official "uplift" programming in the local vernacular blaring from a community receiver that carried no other frequencies. (p. 717)

The hope amongst the colonial British was that rural people would eagerly flock to the radio for education and if not, would eventually come around to seeing its usefulness. After all, this "object was to be used as a weapon against illiteracy and ignorance" amongst the Indian peasants (p. 738). This ideology was deeply embedded in the British psyche to the point where they imposed broadcasting in villages by linking radio programmes with loudspeakers all through the day. There was little escape from colonial altruism. With all this effort, it was found that such programmes proved highly unpopular and such schemes resented by many. The public chose "ignorance" to colonial sponsored enlightenment. It doesn't take much of an imagination to draw parallels to current State policies in scaling of technologies for human development: satellite television for non-formal education (see Perraton, 2000), Geographic Information Systems (GIS) for national planning in agriculture, land use and property ownership (see Hoeschele, 2000), to the mass-scaling of computers for socio-economic mobility.

That said, there are genuine reasons to see new technology as a serious actor in the betterment of human development. These tools come with their own deliverances and promises in the form of new vaccinations and remedies to new media expressions. The green revolution in India, for instance, is said to have expanded India's self-sufficiency in high-yield crops for sustenance of a growing population. Although, some critique this as short-term solutions with long-term damaging ecological consequences (Shiva, 1991b), it doesn't take away from the fact that India's dependency on the West for food imports reduced markedly, contributing to a more confident and productive nation.

In this light, how should we perceive computers for development? It needs to be noted that, while almost 100 percent of the Indian population currently have access to the radio, and about 60 percent have access to the television, less than 5 percent have access to the computer and the Net (Keniston and Kumar, 2004). Yet, the scaling and usage of computers in India in recent years have led to some positive results. Perhaps the most promising use of ICT in India lies in what has been termed "e-governance," involving the computerization of State governance for greater transparency. Connection of the central government to its local counterparts and the provision of online government registrations, pensions, land records, legal proceedings and taxes are beneficial, particularly given the reputation of corruption amongst the Indian bureaucracy. Of course these initiatives are seen as highly cost intensive and contingent on fundamental changes in State processes, for which, there continues to be strong internal resistance.

Some success however, has been achieved. The national digitalizing of passenger reservations by the Indian Railway system, online registrations through the CARD initiative and municipal tax collection online in Andhra Pradesh, online custom declarations, digital land records in Karnataka through their new *Bhoomi* centers and more are seen as having reduced corruption dramatically across these States. In fact, a survey of 4,500 citizens from Hyderabad, Delhi, Mumbai [Bombay], Kolkata [Calcutta], and Chennai [Madras] by the Center for Media Studies shows the following:

> ...that e-governance has brought down corruption in India. The study covered basic services, electricity, municipal corporations, urban development, transport, civil supplies, hospitals, water supply, and railways. Between 2000 and 2004, corruption went down from 63% to 27% in Hyderabad, from 51% to 19% in Kolkata, and from 38% to 18% in Chennai. But the level of corruption stayed about the same in Mumbai and even spurted from 40% to 49% in Delhi. (Pathak and Prasad, 2006, p. 439)

On the other hand, the "i-governance" of interactivity between citizens and governments for strengthening democracy is seen to be far more formidable to implement. This doesn't however stop initiatives from springing up including corporate-sponsored programs on fairer agricultural pricing to more efficient banking practices.

In fact, recent efforts of computer kiosks for socio-economic development yield interesting findings. It is found that people frequent these centers for a range of purposes such as to check local weather updates, express public grievances online, apply for licenses and birth certificates, file petitions, print land records, check examination results and prices of agricultural goods and services (Sreekumar, 2007). Further, there is demand for computer-skills for data entry jobs and distance learning classes. In many cases, "a kiosk brings to the villager information that once was practically unavailable, and services that took days of costly travel are now within biking distance" (Rangaswamy and Toyoma, 2006, p. 5). It is argued that such kiosks could save a village 15,000 dollars annually in costs of trips into town and lost wages. Other kinds of usage are more social in nature such as access and usage of email, chat, horoscopes, video gaming, photo-shopping images, watching videos, downloading music and 'skyping' with relatives overseas. Further, studies have found that "most operators express unqualified satisfaction with their kiosks – even when struggling to break even" (p. 8). Beyond these economic benefits to the entrepreneur, larger social benefits are seen to include strengthening of family relations and gender empowerment through the use of computers (Best and Maier, 2007).

However, there is a dearth of deep analysis in the perception and usage of computers in India. As we have seen above, such investigations are often targeted to explore causal relations between socio-economic benefits and computer usage. Furthermore, such research for the most part is funded by the State and corporations with an apparent conflict of interest when gauging the impact of computers. In this book however, the goal is neither to prove the success or failure of a particular initiative nor measure the socio-economic benefits of computers for the community. Instead, the preoccupation here lies mainly in capturing the range of beliefs and practices with computers amongst a largely undocumented population in Central Himalayas.

PART I
Almora

Chapter 3

This is India, Madam!

My shoes are Japanese   मेरा जूता है जापानी
The pants are from England   पतलून इंग्लि तानी
The red hat on my head is Russian सिर पर लाल टोपी रूसी
But even then, my heart is Indian फिर भी दिल है हिन्दुस्तानी

Famous song lyrics from *Raj Kapoor's* 1955 Bollywood movie *Shri 420*

In Search of a Man-Eating Catfish

Two heavy-duty fishing rods in military-like encasements rest alongside sleeper bunks in a First Class air-conditioned coach on the *Raniketh Express*. This is an eight-hour overnight journey heading from Delhi to Kathgodam station in the Central Himalayas. Two British men, one holding a camcorder while the other one narrates, position themselves alongside these cases on the top bunks, seemingly unaware of us casual spectators. Bogey mates for now, we listen in on the compelling tale. We learn that these gentlemen were in India two years earlier living in a small village a few hours away from Kathgodam where fishermen were mysteriously disappearing near their local lake. It became known that every time someone went to fish, they did not come back. Rumor spread that there was a large man-eating catfish residing in these waters. Our travelers, having won a grant to make a film on this legendary man-eating catfish were returning to prove the rumor. They had come to catch the fish.

Perhaps on a similar journey, I am heading to Almora in search of my own catfish. Rumor has it that villagers are flocking to computers, seeking a pathway to economic nirvana. Rumor has it that villagers are collectively *figuring out* this new technology. Cellphone in one hand and mouse in the other, they are being celebrated as the new "netizens" in the making. The bottom of the pyramid is seen to have flipped. My journey is about catching this rumor and making it known, not so much in proving its falsity or truth but in capturing movements, voices, reason for existence and, perhaps, a little truth. After all, rumor exists because of its grain of credibility.

Armed with such aspirations, I embark on a three-hour taxi drive to Almora town from the train station, a mesmerizing yet nauseous experience. With long and winding pathways up the mountain, you are advised to suck on a lemon or ginger to keep your stomach in check. At six in the morning on a freezing January day in the Himalayas with fog guiding the way and headlights in full swing, you can see little flickering lights along route (see Figure 3.1). Stars in the day time, they happen to be scattered tea stalls equipped with black and white televisions switched on early to catch the cricket test match series between Australia and India. As you turn corners, hoardings of *Airtel, BSNL, Reliance* and *Vodafone* cellphone services flirt within eyesight. There is a large billboard of Shah Rukh Khan, the king of Bollywood or "King Khan" as they say, holding a "1" for *Airtel.* A sign for Birla institute passes me by.

While dodging fog may seem adventurous enough for some, others take it further. Few use seat belts despite the narrow streets and dangerous turns. We stop briefly to get some *chai.* While my driver and I sit at the tea stall before a fire, me on a plastic chair while he insists on standing, a car comes to a halt. A young newlywed couple, as revealed by henna on the woman's hands, joins us by the fire. We start to talk. The woman says nothing. I am deprived even of eye contact. The man, on the other hand, embarks on a questioning spree. He asks me where I come from. I tell him New York. He pries further. His enquiry reaches my street address. It is his way of showing he knows New York well. And in fact he does. He spent a few years as a student there, and simultaneously worked at a cellphone shop at Fashion Avenue with his uncle. He decided to come back to India for a better life, not so much economically as socially, "you can be very alone there…life is short,

Figure 3.1   House on the Hill in Almora

Figure 3.2   View of Settlements in Almora

no point staying unnecessarily alone when I belong to India." In the background, their taxi driver comments that his "mobile is in a coma" (मोबाइल बेहो गि में है). Another jokes that at least it's not BSNL (the government cell phone company). After all, he remarks, it really stands for "even God cannot get through the line" (भगवान से भी नहीं हो सकेगा). People laugh. Our journey resumes.

Naveen, my taxi driver, takes out a package while he drives and asks me to look at it. He says it's a famous herbal product of Almora, which grows only in this Kumaon region of the Himalayas (see Figure 3.2). And like *neem*, a herb reputed to cure many ills, this is to be the next herbal miracle. He claims that his family uses it all the time and even invested in it by getting a 500 rupee share with a Malaysian company that now owns and markets this product. He says that they use a "multi-level marketing" approach through the Net. Surprisingly, he doesn't press me to buy any. As my time in Almora lengthens, I discover this is very much part of the *Pahadi*, or hill people reserve where they distinguish themselves often consciously so from the more "brazen" *Delhiwalla*s (Delhi people) who frequent these mountains. So after this short pitch for this "international" product, Naveen pops in a tape of old *Kishore Kumar* Bollywood hits that carry us through all the way to Almora, "there is nothing like these evergreen songs…look at today's lyrics, they do not make any sense at all" (ऐसे सदाबहार गाने हैं ही नहीं। आजकल के गानों को देखो, बिल्कुल बेतुके हैं) I nod at the back.

As we enter Almora town, the well-known Bright End corner greets us. Made famous by Swami Vivekananda (arguably the first of the Hindu missionaries in the West) and the *Lonely Planet*, it claims to have some of the best sunrises and sunsets in the Himalayan region. In fact, over my stay here, the linkage of famous personalities with Almora becomes a common occurrence, earning its title as the cultural heartbeat of Kumaon. Mahatma Gandhi has reminisced about these hills, "…after having been nearly three weeks in Almora Hills, I am more than ever amazed why our people need go in Europe in search of health" (1971, p. 4). Uma Thurman spent much of her childhood days here. The first British Nobel Laureate (in medicine) Sir Ronald Ross was born here and artistic giants such as D.H Lawrence, Bob Dylan, Cat Stevens and Timothy Leary, the father of the hippie movement, marked these hills with their extended stays. To this day, *Cranks Ridge* or hippie hill, about 20 minutes from Almora town, continues to draw foreigners who not just come to visit but often settle down and intermarry with the locals. So it's not uncommon to see Caucasian-looking youth speaking fluent *Pahadi*, the local hill language, riding their bikes with sari clad women behind. It is also not uncommon to find German pickles, mayonnaise, *Godiva* chocolate, and *Heinz* beans on toast in obscure small shops in this part of the mountains. In fact, one of the top Indian women activists of this region is a product of the hippie days, half Indian and half Belgian. She built her reputation on resuscitating the local women's ancient weaving practices as well as taking on the lumber companies by instituting the preservation of the nearby Binsar forests. She is now in her forties, single, with a teenage son going to boarding school in Coimbatore in the South of India.

This kind of "stardom" however, takes a backseat as you enter the township, navigating through a motley traffic of cows, cars, and carts, sharing space on its narrow lanes. Rows of sweet stalls, Kumauni shawls, vegetable and fruit vendors and, yes, cybercafés, inundate the main Mall Road, with signs for intercolleges and schools scattered about (see Figure 3.3). A long row of jeeps or "shared taxis" sit idle alongside the road while eager drivers hustle to get more passengers. A typical transport option in the hills, more so than the buses, these shared taxis usually are old jeeps re-gutted to accommodate around eight to ten people at the back while squeezing two to three in the front. As we get past the taxi stand, a billboard advertising the *Airhostess Academy* comes to view, showing a young Indian woman wearing a tight black suit, smiling confidently. The *School of Languages* follows, claiming instruction competency in "Urdu, Hindi, Arabic and Computers." Off the main road are famous 200 year old cobbled-stone markets, *Lala bazaar* being the most prominent. Old Kumauni architecture houses pharmacies that sport *Sony* and *Kodak* neon signs. Serving as the shopping paradise of Almora, you can find a veritable collection of odds and ends here from plastic glittered flowers, turmeric, saris, glass bangles, Java programming books, to pirated copies of old and new Hindi, Pahadi and English compilations of movies and songs.

Figure 3.3   Signs of Cybercafés on Entering Almora Town

But why Almora for research, one might ask. Its undeniable beauty and its image as the cultural center of Kumaon for almost 400 years is often an underlying reason that draws scholars to this place. Perched at an altitude of 1,638 meters, Almora is surrounded by fir and pine tree forests and framed by snow clad views of the Himalayas. This ancient hill station, shaped as a horse saddle, is found nestled between the river Kosi and Suyal. Settled by the Chand Rajas in 1560 till the 18th century, it was later taken over by the British during colonial times. It was made into an army and cantonment area and also served as their "Switzerland" getaway during the summers. Thereby, it is not surprising to learn that most property in Almora that have the best views hold anglicized names like "Epsworth" and "Bright End." These are often in legal dispute, as locals battle over dubious ownership rights with missionary entities, both American and British. And here in Almora, Hindus make up the bulk of its population, where predominantly Brahmins and Rajputs prevail, besides a small subset of tribal non-caste people or *Bhotiyas* (Negi, 1995). Interestingly, there is no indigenous trader caste, a fact that sets this culture apart from those of the plains. However, all Uttarakhand castes can and do engage in commerce. In Almora villages, three castes prevail: *thakurs* (noble lords), *brahmins* (priestly caste), and *harijans* (lower caste). While caste doesn't play out in all public settings, it does seem to crop up frequently during meal preparation. For instance, sending children to the same school regardless of caste has not been a problem; the problem, it is said, arises during lunch where children sit separately based on caste affiliation. But times do change. In some villages, you will witness joint cooking in the kitchen and active efforts to promote lower caste women as *balwadi* (preschool) teachers. So behind Almora's celebrity and dominantly "upper caste" status, or as people here say, the land of the "Pants, Joshis, and Pandes" lies a layered reality. According to the 2007 Uttarakhand State Government statistics, 90 percent of Almora's 632,866 population reside in villages. Males number 294,984, and constitute roughly 47 percent of the population and females number 337,882 or 53 percent of the population. Also, there is a migration rate of 60 percent to the neighboring States of Uttar Pradesh and Punjab, primarily due to a high rate of unemployment (Sati and Sati, 2000a). This situation is exacerbated by an average literacy rate of 73 percent (higher than the national average of 59.5 percent), with 89 percent of the males and 60 percent of females marked as literate. Unlike other migratory patterns, contribution through remittances to the economy is marginal, making perhaps 5–7 percent of the income. The rest is drawn from crop cultivation and animal husbandry. 90 percent of the population is engaged in subsistence agriculture, where, it is a commonly known fact that women here are the backbone of this agrarian economy.

   In fact, Uttarakhand is most famous for its environmental activism, the *Chipko* movement being the most well known. Much before the tree huggers of California, *Chipko* meaning literally "to stick" in Hindi was a mass response against logging in the early 1970s, when female peasants went up against the lumber companies by embracing trees. With their forests and thereby their livelihood threatened by contractors of the State Forest Department, this seemingly simple act led to a mass movement that swept across the State and arguably took on a more radical form in this region aligning with the larger demand for a new State. In 2000, the new State of Uttarakhand was formed. Ironically, this mass activism has, for better or worse, led to stringent forest laws in this region, leaving about only 9–11 percent of the land available for irrigation (Gadgil and Guha, 1993). Perhaps this strong tie to the land can be partly explained by the unique community and individual land rights in this region predating the British occupation, where most families owned land and had a common stake and voice in forest keeping and access to resources for their livelihood.

So, besides access to limited resources in the land of plenty, villagers in Almora struggle with certain basics for survival, including access to quality healthcare, clean water, electricity, good education, and regular transportation, particularly as households are scattered across this mountainous region. For instance, 55 percent of the villagers have to walk more than five kilometers to reach the nearest river bank in Almora and women spend an average of 4–6 hours collecting fodder and fuel for their day-to-day living (Agarwal, 1992). Also, 60 percent of the rural population lives in areas that are more than five kilometers from the town, where access to most of the markets, hospitals, colleges and other services reside. Yet 70 percent of people make the effort to visit the town at least once per month for a range of services, including medical, pension, employment and education, to name a few.

All this makes Almora a fascinating study, especially in terms of my interest in social usage of new technology, particularly with computers. After all, Uttarakhand, having become a new State, has jumped on the digital bandwagon, making all the right noises to gain the title of an "aspiring leader" in "e-readiness." It has placed itself as an active partner in the formation of India's Silicon Valley, committing to create connectivity across its terrain through a steady supply of computers to high schools, universities, government agencies and public kiosks. It has embraced the not so humble national ICT Indian *Mission 2007* policy that aims to connect 600,000 villages across India through a network of hardware and software. So, currently in Almora, there is a flurry of activity reported from digitalizing data across government agencies, such as land records, pensions, employment opportunities, ration cards, voting registration and more. All government high schools and universities are supposed to have been connected with computers, while broadband and wireless services have just made their way into town little more than a year ago. If we are to understand relatively new technology users and their activity, what better way than to select a site where the ground is ripe for such momentum, where potential and need intermix.

Here Comes Sonia Gandhi!

There is no pretending I am not from here. Knowing Hindi does help but Hindi with an accent after years of city "corruption," both in India and the United States, has situated me well in the often mocked category of the NRI (Non-Resident Indian). But I get a step ahead. With my Convent school accent on my Hindi, I seem to provide entertainment across the board, especially amongst the villagers. I find that, instead of my status as the outsider – the "foreigner" being an obstacle in fieldwork, it often serves as a much needed ice-breaker. Many compared me to Sonia Gandhi, the Italian-born wife of Rajiv Gandhi, and current President of the Indian National Congress Party; she speaks Hindi but with a thick Italian accent; "you talk like Sonia Gandhi but you should not worry, look where it got her!" (तुम सोनिया गांधी की तरह बात करती हो लेकिन तुम्हें परे ान नहीं होना चाहिए, देखो उसने उसे कहां पहुंचा दिया) they joked. I intrigued them.

Women wanted to know why I did not have children, whether I had my own farmland in my New York "village" and why I wore less jewelry; girls wanted to know where had I traveled to, had I ever worn jeans and what did my parents think of my leaving home unmarried. Men rarely asked questions but loved to talk: they spoke of politics, the government, the winter in Almora, the lack of rain, their crops, the produce at the markets to the *Bihari's* coming and taking their jobs. There was no dearth of topics. I found that in general, it took little to get people to talk as long as you met them in spaces of comfort; girls when grouped together away from adults, and women often when away from their men. Men, on the other hand, seemed most receptive to conversation at tea stalls and more formally through appointments.

Swami Mafia

"Swamis [Hindu gurus] are big business. There is this swami, he's a young boy of say 19, maybe 22…people come to listen to him and donate a lot of money to him. Recently they donated 20 lakhs [about 40,000 dollars] to the *Arya Samaj* and other groups who have been supporting this kind of spiritual guidance and then we find all kinds of accusations about this boy…now he's gone and the money's gone too," relates Dhiraj with two gunmen by his side. This man claims to be a simple musician. Over time, the story unfolds: of him witnessing a day time murder in Haldwani in the plains about two hours away from Almora; of him testifying and being threatened by goons with ties to Dubai and Mumbai; of him being given police protection by the government for two years. He claims to be a real estate agent as well. And not to forget politics where he holds some official title in his district. Besides these side ventures, he actively organizes spiritual talks by swamis for the locals, paying for all the hosting expenses.

During my visit, he often came to Almora in search of spiritual guidance in the hands of a swami residing as a guest at *Epsworth*, a colonial home in *Kasar Devi*, occupied by a retired sugar baron from Lucknow. The swami, who went by just "swamiji," came with a Ph.D in "Divine Love" from Germany. He had spent 16 years out there and six in Ontario as a professor in philosophy, before moving to the Himalayas. He came back, he says, to lend guidance to villagers in Almora, to make their lives better. He submitted a proposal to the district magistrate of Almora requesting funds for computers for the surrounding villages where his ashram was being built. His devotee, a confirmed bachelor in his fifties, had worked as a chartered accountant for 30 years in Calcutta. Having heard swamiji's lectures, he decided to channel his life savings into the building of the ashram, packing bag and baggage to follow swamiji to Almora. The ashram, however, was hardly far from controversy. Having run out of money, the ghost of the idea continued to manifest itself in cement and steel remnants in the nearby *Sheehat* village. Our swamiji had lost much credibility. The villagers from there cursed him and any kind of association with swamiji became a risky business. It was said that he took money from the villagers, promising them better health and education, computers and English, a good future for their children. Five years later, the wait turned to anger.

Ten kilometers away from Almora town lay another hub of spirituality. A Shangri-La to some, the members of the *Mirtola* ashram had succeeded in converting a declared barren land into one of high productivity. It was founded by Sri Yashoda Ma, a housewife turned ascetic along with her disciple Gopal Da, a Cambridge educated English fighter pilot in the First World War. Having passed away in 1965, Gopal Da was followed by his disciple Sri Madhav Ashish or Ashish Da, coincidentally a Scottish aircraft engineer who came to India during the Second World War in 1942. Seduced by the ashram ways, he stayed behind, attracting a new slew of disciples from all over India, with what people describe as his pure energy and charisma. During his time, an engagement with the local communities transpired. This involved substantive work in the field of environment and education, for which he was later awarded the *Padma Shri* in 1992 by the Indian government, one of the highest honors bestowed to an individual for ones' contribution to the country. When he died in 1997, his disciple David or Dev Da, credited with pioneering innovative farming and water preservation techniques in the local area, took over.

But why linger on this subject of spirituality? Out here, it is veritably impossible to separate the "spiritual" from the "practical." In fact, every spiritual hub feeds on the non-spiritual to validate itself. Their survival is based on the need to be needed. Be it retired generals, school principals, bank managers, politicians, activists to the local shop keeper and pharmacist, it is not a leap to state that most people here claim a belief in something…somebody. Between the ashrams, swamis, sadhus and the numerous temples that mark the mountain scape, there is no shortage of sites of introspection, meditation and contemplation. While this is hardly an attempt to reify the age-old stereotype of the Indian mystic in the common man, it is more an effort to remind the reader that "spirituality" finds its way in the most unexpected of encounters, strategically perhaps, evoking, eliciting, engaging to justify, explain and share ones purpose. After all, escape seems elusive from the Himalayan brand of spirituality.

And so we notice: in the simple speech of a private high school principal and devotee of the *Sri Sri Ravi Shankar Art of Living* Foundation:

We are chasing pots and pots of money, making plenty of money and always moving, not thinking of anything or anyone else. We look at everybody but we forget to look at ourselves…we should love ourselves. We keep saying *I love you*, we write it, we speak it but do we really live it? Life is like a carrot…we can make a vegetable dish out of it but can we make a dessert from it too? That is life – we are living it but can we enjoy it?

हम रुपयों के पीछे भाग रहे हैं, खूब रुपये कमा रहे हैं और हमे ा भाग रहे हैं, कुछ और या किसी और के बारे में नहीं सोच रहे हैं। हम हरेक को देखते हैं, लेकिन खुद अपने को देखना भूल जाते हैं। हमें स्वयं अपने को प्रेम करना चाहिये । हम आई लव यू कहते रहते हैं, इसे लिखते और बोलते हैं, लेकिन क्या हम इसे वाकई जीते हैं? ज़िंदगी एक गाजर की तरह है, हम इसकी सब्जी बना सकते हैं? कि लेकिन क्या हम इससे गाजर का हलवा बना सकते हैं कि नहीं? यही ज़िंदगी के बारे में है, हम इसे जी रहे हैं, लेकिन क्या हम इसका मज़ा ले सकते हैं?

She sends her teachers as part of their teacher training to *Art of Living* seminars; she gives out brochures to parents in the hope that they would visit the Foundation and take part in these workshops; she is in constant process of cajoling, converting, and convincing.

Then there is the retired *Arthur Andersen* Brahmin consultant from Mumbai with ancestral ties to Almora. He and his wife moved to Almora a few years ago, having built a three story home off Bright End corner, prime property with prime views. While his wife tends to the garden and misses her friends in Mumbai, he reads the Upanishads, seeking historical insights into Brahmanism, caste, life. He sends out emails with links to articles from the *New York Times*. He cites Nicholas Dirks on the caste system. He no longer works for profit. He was warned by a swami that whatever he now touches with commercial intent will fail. Living simply and thinking deeply is his prime goal now. He calls himself the "one dollar a day man." Local people call him "the man with the big house" (बड़े मकान वाला आदमी). In search of guidance on the Upanishads, he sought swamiji and in exchange, drafted the proposal on behalf of swamiji for rural development in Almora.

And what about the young newlywed couple with their NGO? Having failed to get into the prestigious IAS (Indian Administrative Service), Mohan, the husband, was fortunate to get onto a UN sponsored project on water management in Almora. Following this, he founded an NGO a few years ago and is in a constant struggle to find funding, given the burgeoning of NGOs in this area. In fact, with the formation of the new State, a flow of cash has come in from the national government in the name of development, resulting in this area having one of the highest concentrations of NGOs (Sivaramakrishnan and Agrawal, 2003). So Mohan gets ready to be equally passionate about women's empowerment, environmental degradation to rural ICT kiosks for villagers, depending on where the flow takes him. While there are currently no government ICT kiosks for villagers in Almora, there are plans, he says, to start them. The government has expressed an interest in a proposal for an all-in-one, one-stop-shop ICT kiosk for government services in villages in Almora. Mohan intends to be first in line for this job.

Meantime, passion does not wait. Having gathered the villagers on the issue of water harvesting and the new government policy of water catchment areas, Mohan makes his speech before the villagers:

The question is, who needs to be treated? Man or Water? You treat man and water will follow. If we are cold, we cut trees for fuel, but did we think of planting back the trees? So even if you cultivate the small landholdings, the land will be poor if others around you do not treat the land as well.

सवाल है कि किसका उपचार करने की जरूरत है? इंसान का कि पानी का? इंसान का उपचार कर दो पानी का अपने आप हो जायेगा। अगर हमको ठंड लगती है तो हम ईंधन के लिए पेड़ों को काट देते हैं, लेकिन क्या हमने पेड़ लगाने के बारे में सोचा? इसलिए अगर आप छोटी का तों में खेती करते हैं तो जब तक आसंदकपास के लोग भी भूमि को न सुधारें भूमि घटिया रहेगी।

Back to spirituality: it is worth noting that while it may seem that at every turn is carved an ashram of sorts, it is more a practice amongst a certain class of people: those urbanites in search of peace of mind, beauty, mindful reflection; foreigners seeking to stretch the magical sixties, and find alternative social models to live by, even if for a brief stay; Israeli soldiers taking time to deal with their internal conflict; tourists curious about this saffron clad culture with hidden cameras in hand. As one shopkeeper who sells his wares to one of the ashrams remarks, "if you're a swami, you're probably not from here. If you're a sadhu, you'll probably go to Rishikesh" (अगर आप एक स्वामी हो तो भायद आप यहां के नहीं हो। अगर आप एक साधु हो तो आप भायद ऋषिके ा जायेंगे). Imported swamis come with wealth, while the poor sadhus head to the sacred centers of Rishikesh and Haridwar, relying on local mass support. With money comes the need to set root, take land and make it theirs. And ashrams need space to appear serene to attract further funding.

Privatizing scarce land does not go unnoticed though. To appease local villagers, ashrams often provide free basic health services and employment to the lucky few. Ashrams are not meant to discriminate amongst people; poor or rich likewise are to enjoy the benefits. Yet, gates to ashrams are often locked with heavy security at their entrances. Besides this physical barrier, what keeps the villager at bay? As one villager responds, "who has the time to meditate? If we meditate, we will be thinking of our hut, our cows, our children, our field...the mind will be having a busy signal every time" (ध्यान करने का वक्त किसके पास है? अगर हम ध्यान करे तो, हम लोग अपनेंदकअपने झोपड़ी, अपनी गाय, अपने बच्चे, अपने खेतों के बारे में सोचते रहेंगे...हमारे दिमाग को हर समय व्यस्तता का संदे ा मिलता रहेगा।). Another states frankly, "people basically want jobs or free hospital services but everyone knows that these ashrams buy so much land and need to keep us happy so they start some basic hospital. They just want to buy our goodwill so we don't make a scene" (लोग मूलतः नौकरी चाहते हैं या मुफ्त का इलाज लेकिन हर एक यह जानता है कि यह आश्रम इतनी ज्यादा जमीन खरीदते हैं और हम लोगों को खु ा रखते हैं ताकि वे मूलतः एक अस्पताल भारू करें। वे सिर्फ लोगों की गुडविल खरीदना चाहते है ताकि हम लोग कोई बखेड़ा न खड़ा करें।). If so, then where do most local people worship?

Ritualistic for the most part, they celebrate life events and festivals throughout the year at home, as well as the temple, through a combination of prayer, food, and social gatherings. Almora has no shortage of temples and, in fact, lays claim to the title of the "city of temples," dating back to the Chand rulers of the 1600s (Negi, 1995). Dedicated to forms of Shiva, Durga, Ganesha or Vishnu, these temples attract a mass of worshippers for a range of purposes all through the year. The famous *Chitai* temple is revealing: sacrifices and promises are made here. A narrow entrance framed by bells leads to the main temple, leaving behind the commerce of religion in the guise of *prasad* (temple offerings), red and gold threads, deep fried *pakodas* and *chow mien*, a reminder of Almora's proximity to China.

The temple has gained the reputation of a wishing well of sorts. Notes tied to bells at this site reveal the yearnings of many: "if you make my husband better, I promise to donate money and sacrifice my services to you," promises one; "I hope my parents will live long and happily and that I keep my job;" "my court case is still going on and there does not seem to be any justice coming my way. If you give me my justice, I will donate 50,000 rupees [1,000 dollars] to you," bargains another. Most common requests get scribed again and again to these bells: to get into the prestigious Indian Administrative Services (IAS), to pass the school exam, to wish for the health of a loved one, to pray for their husband to come back from the army, to ask for their mother-in-law to be nice to them, to hope for a quick settlement of a legal dispute…a glimpse of this society all rolled up in sacred gold and red thread, leaves their messages ringing for everyone to listen.

That said, in the following chapters, particularly when dealing with computers, we see little of "spirituality" seeping through in events and discourses in spite of its argued omnipresence. So although swamis have evoked and facilitated some initiatives to disseminate and use computers for rural development, the actual usage and understanding of computers on a day-to-day level is seen to have little to do with one's spiritual orientation. But rather than disregard such spiritual leanings, perhaps we can see it as reason to keep as a backdrop when encountering interactions with computers. In doing so, the very absence of spirituality in discursive practice about computers in the following chapters is itself a statement on the meaning of computers as constructed by people in Almora.

In God We Trust, the Rest is All Cash: The Simple Villager?

Taking comfort in the "simple" villager syndrome has been an age old occupation and hobby to some. Synonymous to a "primitive" being, the villager has been "frozen" in time (Appadurai, 1988) and laced with notions of empathy, solidarity, collectivity and community. Contemporary rural-urban discourse three-dimensionalizes such settings, an expected pendulum shift in the world of scholarship. Yet what are we to make of these enduring relationships to land…to ceremony? As a well versed ethnographer with hope for difference, I encounter similarity: semblances to colonial snapshots, of past documentations, of urban circulations within rurality. Yet, on looking closer or at the blink of an eye, the image seems to blur.

Posted outside many *paan* or beetle leaf stalls in Almora, the sign reads, "In God We Trust, the Rest is All Cash" (हम लोग भगवान में वि वास करते है बाकि सब नकद है।). Prema, a lower caste woman from Hawalbagh village remarks, "shopkeepers never forget…they give on credit but they always take back, even if its months later" (दुकानदार कभी नहीं भूलते। वे उधार देते हैं, लेकिन वे हमे ा वसूल लेते हैं, चाहे महीनों के बाद ही सही). Money is secular: rural or urban, upper or lower caste, local or foreigner, the tab goes on. Trust is an interesting phenomenon. In the village, shopkeepers give credit without proof of assets, or known affiliation. This is the most rational of practices though. This is good business sense. Trust is tied to its remoteness and dependency, given the skewed ratio of few consumption outlets and many consumers. Prema needs to go back for more. She takes credit for a passport photo for her ration card, her daughter's photo for the board exam, her sister's marriage album.

Prema has never left Almora. Chandan, her neighbor, on the other hand, left Almora when he was a mere boy of 15 to work as a servant in Lucknow city. Eleven years later and with a four year stint in the capital, Delhi, where he drove a taxi cab, he made his way back home. "Why did you come back?" I asked. "It's a long story," he says, going ahead in his narration:

We used to be a joint family…there were 75 of us on the same land…can you imagine…three women just to make the *rotis* [bread], four women washing dishes outside the house, the older women supervising… it looked like we were running a large restaurant (laughs). But then these women have become all modern…they have become very demanding. One woman sees what the other women get in the family and demand it from their husbands. The poor guy, whether he can afford it or not he will have to get it for her otherwise she will threaten to leave for her *maike* [family home]. So the man even borrows money just to make her happy. One day it's a new petticoat, another day a new sari blouse…it never ends. And then they do *kus pus* [slang for gossip] into their husband's ears and next thing you know all the brothers are fighting amongst themselves. So my grandfather decided enough is enough. He said to us all to split the

हमारा संयुक्त परिवार हुआ करता था, हम 75 लोग उसी जमीन से खाते थे, आप कल्पना कर सकती हैं! तीन औरतें केवल रोटियां बनाने, चार औरतें घर के बाहर प्लेटें धोने, उम्रदराज़ औरतें इन कामों पर नज़र रखने के लिए; ऐसा लगता मानो हम एक बड़ा रेस्टोरेण्ट चला रहे हों (हंसते हुए)। लेकिन अब ये औरतें पूरी मॉडर्न बन गयी हैं, उन्हें बहुत कुछ चाहिये। औरत देखती है परिवार की दूसरी औरतों को क्या मिला और अपने पतियों से उसी चीज़ की मांग करती है। बेचारा, वह चाहे समर्थ हो या नहीं, उसे वह चीज़ उसके लिए लानी ही होगी नहीं तोपत्नी अपने मायके चली जाने की धमकी दे देगी। इसलिए सिर्फ उसे खु ा रखने को आदमी रुपये उधार लेता है। एक दिन नया पेटीकोट, दूसरे दिन नई साड़ी ब्लाउज; यह (सिलसिला) कभी खत्म नहीं होता। और फिर वे अपने पतियों के कानों में खुसुरंदकपुसुर करती हैं और इसके बाद आपको दिखता है कि भाई आपस में ही लड़ रहे हैं। इसलिए मेरे दादा ने फैसला किया कियह बहुत हो चुका। उन्होंने हम सबको जमीन की हिस्सेबांट करके भांतिपूर्वक रहने को कहा। अब मुझे वापस आना पड़ा, क्योंकि घर में मेरे बच्चों और पत्नी की देखभाल करने वाला कोई नहीं है। मेरे पिताजी अब बहुत ज्यादा बूढ़े हो गये हैं इसलिए घर पर कोई पुरु ा नहीं है। और आपको पता है कि ऐसा

land and live separately in peace. Now I had to come back because there is no one in the house to take care of the children and wife. My father is too old now so there is no male at home. And you know how it is, the house is a full half hour off the road in the jungle. But before I did not want to leave Delhi and now I don't want to leave Almora.

कैसे है, सड़क से मकान तक पूरे आधे घंटे का जंगल का रास्ता है। पहले मैं दिल्ली नहीं छोड़ना चाहता था और अब मैं अल्मोड़ा नहीं छोड़ना चाहता।

Women here are seen as strong. They have built a reputation for endurance, toughness, and bossiness. The woman of the hills, men readily acknowledge, does most of the work while, "we men just play cards and drink *chai*" (हम आदमी सिर्फ चाय पीते हैं तथा ता ा खेलते हैं।). And let's not forget alcohol. Male consumption of alcohol is perhaps the single most common issue that perpetually circulates in a woman's complaint, manifesting in posters stating, "Bottle or Bread?" (बोतल या आटा) by local NGOs and the government. After all, alcohol consumption is often tied to domestic violence and deep debt. Such behavior, sadly a documented common rural phenomenon, is for the most part attributed to the high rate of unemployment in rural areas (Narasimhan, 1999; Samanta, 1999; Sethi, 1999).

Besides this, you rarely hear women complain; not about the fact that it takes them an average of three to four hours to fetch water, which they carry back home on their heads through the mountain trails. They seldom draw attention to their collecting of fuel and fodder, looking after their children, domestic chores and their tending to the cows and more (see Figure 3.4). For tilling a one hectare farm, it is documented that women put in 640 hours for weeding, 384 hours for irrigation, 650 hours for transporting organic manure to the field, 557 hours for seed sowing (with men), and 984 hours into harvesting and threshing (Shiva, 1991a, p. 5). In spite of this intensive labor, women continue to experience lower status in the mountains.

A female child and her grandmother attend a meeting organized by a local NGO on getting women to run for elections for the *panchayat*, the village political body. The government has just passed a law requiring 50 percent of *panchayat* positions to be occupied by women, but women are not coming forth. The child is about eight years old. She sits in the corner, staring blankly at the wall for the entire hour. Some women from the NGO start to get concerned. They are convinced that she is sick and try to persuade the grandmother to take her to the hospital. After much pressure, the girl is taken to the local doctor. She is diagnosed with diarrhea and severe malnutrition. The story comes out. It is said that when the mother gave birth to the girl, she was extremely upset. She decided not to breastfeed her. When the baby survived the first two days without milk, people in the village beseeched the mother to feed the child. Eight years later, the girl suffers her fate of being born a girl.

Figure 3.4   Woman and Girl Carrying Firewood

But change happens, albeit slowly. Champa, a well respected and much loved social worker has, for the last 15 years, been channeling her energy in training teenage girls to be *balwadi* or pre-school teachers in their villages. Champa herself was a *balwadi* teacher through the training of a local NGO and, 25 years later, passes on her depth of knowledge to eager yet scared village girls, unsure of change. She is unmarried and lives with another social worker in her forties too, also unmarried.

Champa is on a constant lookout for young girls with "potential," she says. She knows a leader when she sees one. She also knows how difficult it is to convince the family to allow for the girls' training. But years later, she has become savvy in the art of persuasion:

I one day visited this village and saw this girl Aarti. She seemed so capable and strong minded so I thought she would be very good for our work in the *balwadis*. So I asked her if she would come for training to town for ten days. The NGO would provide for everything – transport, food and board and take good care of her. She laughed. She said she could never do that as it was too far and besides, her mother would never let her go. So I said, fine I'll talk to your mother. When I said that, she got scared. She said, no you can't because my mother is in the jungle collecting wood. So I said, okay, I'll wait. While I waited, I told her that she should not be so scared about the outside world. Look at me, I said. I roam alone and so why should you be so afraid all the time? When her mother came, we went to the house and she made some tea and that's when I chatted with her. I asked her if she wanted her daughter to be scared all the time. Should she not have some freedom, I asked? Don't you want what you weren't able to have in your youth? The mother said why, what's the point for Aarti will get married soon and have babies then there's no question of her leaving her home for these things. I asked is that all there is in a woman's life? And so I talked and talked and tried to convince her. Maybe

एक दिन मैं उस गांव में गयी जहां इस लड़की, आरती को देखा। वह बहुत समर्थ और मन की पक्की लगी इसलिए मैंने सोचा कि वह हमारी बालवाड़ी के काम के लिए बहुत अच्छी रहेगी। मैंने उससे पूछा कि क्या वह 10 दिनों के प्रशिक्षण के लिए बाहर आ सकेगी। संस्था आने तक जाने, भोजन और रहने की व्यवस्था करेगी और उसका अच्छी तरह ख्याल रखा जायेगा। वह हंसी। उसने कहा वह ऐसा कभी नहीं कर सकेगी, क्योंकि बहुत दूर जाना है और इसके अलावा उसकी मां उसे कभी नहीं जाने देगी। इसलिए मैंने कहा, अच्छा मैं तुम्हारी मां से बात करूंगी। जब मैंने यह कहा तो वह डर गयी। उसने कहा, नहीं आप ऐसा नहीं कर सकतीं, क्योंकि मेरी मां लकड़ी इकट्ठा करने जंगल गयी है। मैंने कहा, ठीक है, मैं इंतजार करूंगी। जब मैं इंतजार कर रही थी, मैंने उससे कहा कि उसे बाहरी दुनिया के बारे में इतना डरना नहीं चाहिये। मुझे देखो, मैंने कहा, मैं अकेली इधर से उधर जाती हूं इसलिए तुम्हें हमेशा इतना ज्यादा क्यों डरना चाहिये? जब उसकी मां आयी तो हम मकान में गये और उन्होंने चाय बनायी और मैंने उनके साथ गपशप की। मैंने उनसे पूछा कि क्या वह चाहती हैं कि उनकी बेटी हमेशा डरपोक बनी रहे। मैंने पूछा, क्या उसे कुछ स्वतंत्रता नहीं होनी चाहिये? क्या आप वह नहीं पाना चाहतीं जो अपनी जवानी के दिनों में नहीं पा सकीं? मां ने कहा, क्यों नहीं, पर बात यह है कि आरती की जल्दी ही भादी हो जायेगी और बच्चे होंगे, तब इन कामों के लिए उसके घर छोड़ने का प्रश्न ही नहीं उठता। मैंने पूछा, क्या एक औरत की ज़िंदगी में यही सब होता है? इस तरह मैं बोलती ही गयी और उसे मनाने की कोशिश की। शायद मुझे शांत करने को उसने कहा कि मुझे कोई आपत्ति नहीं है, लेकिन इसके पिताजी आपत्ति करेंगे। मैंने कहा, ठीक है, मैं उनसे बात करूंगी। उन्होंने कहा कि वह दिन भर के

to keep me quiet, she said she would not mind but it's the father who would object. So I said, fine. I will talk to him. They said that he had gone for the day for labor. I asked Aarti to take me to where he works. By then Aarti was inspired and said to me that yes, she does want to go to the town and so she took me. When I met the father, I said, see, if you have any concerns of her safety, why don't you come yourself with Aarti for the training and we will pay for your trip and visit and then if you don't like it, you can go back with her. And so that is how they came to Almora town. And after two days of seeing what we were doing, the father left and went back, leaving Aarti behind to complete the training. Now not only Aarti comes regularly but even her mother sometimes to participate in the women's group at the NGO. I think she just likes coming to town for a break.

लिए मजदूरी को गये हैं। मैंने आरती से कहा कि जहां वह काम करते हैं मुझे वहां लेकर चलो। तब तक आरती की प्रेरणा जग गयी थी। उसने मुझसे कहा कि वह भाहर जाना चाहती है, इसीलिए वह मुझे वहां लेकर गयी। जब मैं पिताजी से मिली, मैंने उनसे कहा कि अगर आपको आरती की सुरक्षा को लेकर चिंता है तो उसे प्रि क्षण के लिए अपने साथ लेकर क्यों नहीं आ जाते? हम आपकी यात्रा का खर्च उठायेंगे। वहां आकर देखिए, अगर आपको वहां ठीक न लगे तो उसे अपने साथ वापस लेकर जा सकते हैं। तो इस तरह वे अल्मोड़ा भाहर आये। पिताजी ने दो दिनों तक देखा कि हम क्यांदकक्या कर रहे हैं, तसल्ली होने के बाद वे आरती को प्रि क्षण ा पूरा करने छोड़कर वापस लौट गये। अब न सिर्फ आरती नियमित रूप से आती है, बल्कि कभींदककभी उसकी मां भी संस्था में आयोजित महिला संगठनों की बैठकों में भाग लेने आती है। मैं सोचती हूं वह (रोज़मर्रा की ज़िंदगी से) अवका ा के लिए भाहर आना पसंद करती है।

Balwadis in fact are more than just centers for learning. They are most popular in villages as mothers can drop their children here while on their way to the forest to collect firewood. These centers also conduct health checks and government medical immunization. In fact, this is a preferred venue for some NGOs who are able to push their agendas, given that these centers are one of the few spaces where women routinely get together. Hence, these spaces serve as a multi-platform for "development," creating awareness on women's rights, local politics, education and employment. So Champa not just works on training the girls to be good *balwadi* teachers, she also attempts to get these girls to become more self-sufficient and savvy in ways of life:

I talked to the girls about saving and they laughed. They earn 700 rupees per month from the *balwadis* so I told them why don't you get a bank card and deposit it in the bank. They said that there were no savings left…they had to spend all their money. I was surprised. I

मैंने लड़कियों से बचत के बारे में पूछा तो वे हंसी। वे बालवाड़ी से 700 रुपये मासिक कमाती हैं इसलिए मैंने उनसे पूछा कि वे बैंक में खाता खुलाकर इसे वहां जमा क्यों नहीं करतीं। उन्होंने कहा कि कुछ नहीं बचता, सारे रुपये खर्च हो जाते हैं। मुझे आ चर्य हुआ। मैंने पूछा किस पर? उन्होंने कहा, दीदी आजकल सलवार सूट बहुत मंहगे हैं। हमने

asked on what? They said *didi* [big sister], salwar suits costs a lot these days. So we did the math. I asked them how many suits they needed to buy and they said four a year so I said that with 150 rupees per suit, it comes up to 600 rupees plus 100 for stitching. So they still had a good amount they could save. Then they said very openly that perhaps I did not wear lipstick and dress up but they needed to and so all their money goes into these things like shoes and lipsticks and bangles. So I said okay. But still you should save something and learn to use the bank at least. They said it's too far – at least three kilometers or more. I said that you walk more every day for water so why can't you do this once a month? I told them that when they get married and go off to Delhi, then their husband will have to take a day off from work only because these girls did not learn to go to the bank. And by then you will become too afraid to do so I said. Besides that, if they don't learn to deal with money I warned them that their mothers-in-law will have full control of the house. This scared them. They then admitted that yes, going to the bank is not a bad idea at all.

गणित लगायी। मैंने पूछा कि उन्हें कितनी सूट खरीदनी पड़ती हैं तो उन्होंने जवाब दिया कि साल भर में चार। मैंने कहा कि 150 रुपये प्रति सूट के हिसाब से यह छ: सौ रुपये हुआ, जिसमें 100 रुपये सिलाई के जोड़ देते हैं। इस तरह तब भी उनके पास काफी रुपये बच सकते थे। तब उन्होंने बहुत खुलकर कहा कि भायद मुझे न हो लेकिन उनको लिपिस्टिक लगाने और अच्छेंदकअच्छे कपड़े पहनने की जरुरत होती है और उनके सारे रुपये ऐसी ही चीजों, जैसे जूते, लिपिस्टिक और चूड़ियों में खर्च हो जाते हैं। मैंने कहा ठीक है। लेकिन फिर भी तुम्हें कुछ बचत करनी चाहिये और कम से कम बैंक का उपयोग कैसे करें इसे सीखना चाहिये। उन्होंने कहा कि यह बहुत दूर, 3 किलोमीटर या इससे अधिक है। मैंने कहा कि तुम पानी के लिये रोज़ाना इससे अधिक चलती हो तो महीने में एक बार क्यों नहीं कर सकतीं? मैंने उन्हें बताया कि जब उनकी भादी होगी और वे दिल्ली जायेंगी तब उनके पति को सिर्फ इसलिये काम से एक दिन की छुट्टी लेनी होगी कि इन लड़कियों ने बैंक जाना नहीं सीखा। तब इसे करने में बहुत डर भी लगेगा। इसके अलावा मैंने उन्हें चेताया कि यदि वे रुपयों का लेनदेन करना नहीं सीखेंगी तो घर का सारा नियंत्रण ा उनकी सास के हाथों में रहेगा। इस बात से वे डर गयीं । तब उन्होंने स्वीकार कर लिया कि बैंक में जाने का विचार बिल्कुल बुरा नहीं है।

These narratives tell us something of the shifting dynamics of village life, commerce and consumption, and of the intricate relationship between consumption and gender, contradiction co-existing. It compels us to loosen our grasp over popular concepts of women as "natural" savers and women as victims. In fact, hints of migration and mobility seep through this work. Larger hints of politics gain transparency, between fathers and daughters, of mothers-in-law and daughters-in-law, of shopkeepers and customers, and of NGO workers and villagers. The stability of village life is often teased with. Village girls sporting lipstick and heels to attract suitors are juxtaposed against arranged marriage to nice well-off boys who have made it in the city. Champa, envied for her independence and freedom, is also pitied for her lack of marital prospects. The father who allows his daughter to go to town is blocked by his mother who condemns this as weakness.

Social values don't change as easily, says Dr Harshad Prasad Pant, a septuagenarian with 11 college degrees behind him. A renaissance man of Almora, he has built a reputation of trust through 45 years of service as a surgeon in the district hospital in Almora. He now has set up his own practice at home, giving free health consultations. Strangely though, instead of lines outside his door, you witness a trickle of people coming by. He has built a reputation of eccentricity; as one jokes, "once his patient goes in with a minor illness, they come out with a major headache" (यदि एक बार उसका रोगी एक छोटी सी बीमारी के लिए अन्दर जाये, तो वे बड़ा सिरदर्द लेकर बाहर आते हैं।). His treatment includes a unique dosage of Ayurvedic remedies and lectures on history, poetry and stories of the past. "I just want an injection but instead I get the *Bhagwat Gita*, [holy book]" (मुझे सिर्फ एक इन्जेक्शन चाहिये था परन्तु मुझे भागवत् गीता मिली (पवित्र किताब)) laments another. Villagers complain about their inability to afford the time for long-term remedies as they need to get back to work as soon as possible. "If I don't feed the cows and get the firewood, who will?" (यदि मैं गायों को चारा न खिलाऊं और लकड़ी न लाऊं तो और कौन करेगा?) says a young mother of three; "my daily wages barely covers us...with five children to feed, I had to get my daughter to quit school to work with me on the road construction work...so who has the time to be sick?" (मेरा दैनिक वेतन हमलोगों के लिए है हमारे पांच बच्चों को खिलाने के लिए है, मुझे अपनी बेटी को स्कूल छुड़वाना पड़ा ताकि व मेरे साथ सड़क निर्माण कार्य में मेरे साथ काम कर सके...किसके पास वक्त है बीमार पड़ने का) says a father and *Bihari* laborer who has immigrated here in search of work.

But nobody disputes Dr Pant's knowledge and love for Almora, its history, its people. Reluctant respect follows. Dr Pant admits that his ways have driven people away. He claims that people here want a quick fix for everything. Villagers, he believes, have come to associate injections and antibiotics with any illness. They want action. Even if there is no cure, they want it all. They have gotten into a habit of mind, he says, of expecting such treatment because all other doctors do the same – if it's a headache, give them an MRI, if it's a stomach pain, an ultrasound, if it's a cold, an x-ray. He calls this a money-making racket and villagers buy into it. And painkillers are their favorite. The pharmacist is their god, he says wryly.

Of course with education he says, things change but social values don't change easily, "even Aishwarya Rai [a famous Bollywood actress] still keeps *karva chaut* [a woman fasts for the long life of her husband]!" he remarks. Besides, this area having such a high literacy rate, he argues, is an "optical illusion...they conveniently omit women over 60 and focus more on the town people but barely touch the villagers." With women he says, where is the time for education? There is only "production and reproduction." There is no rest for them. "There is an old *Pahadi* saying," he narrates, "women are like bananas, if you don't consume them quickly they will spoil; men are like apricots, the longer you keep them, the better they get."

So, be it on issues of health, education, or early marriage, much of what we have seen on the plight of women in rural areas has been widely documented. We have gone beyond the "victimology" discourse into that of "empowerment," at times romantically so, with current development practice crowning women as masters of their own destiny (Mohanty, 2003). Feminist scholars debunk historical and common notions of women's role in survival. Some claim that 50 percent of hunting and gathering historically has been done by women, where it is argued, the crux for survival rested more on gathering; "Man is basically a parasite – not a producer," argues Mies (1986, p. 251), turning the patriarchal paradigm on its head. From viewing women as helpless victims and beneficiaries to transferring all responsibility onto their shoulders, development gets caught in its own web.

Meantime we live with a live paradox, perhaps best revealed in this story of a city nurse who has been working with villagers for the last 12 years. When her husband died and her children grew up and left for college, she found herself alone. She decided to spend the rest of her life in the mountains serving the villagers. She came with enthusiasm and new ideas to engage, educate and build awareness. Years later, she chuckles on her naïve days:

> Once we went to this village and during the health training, I thought it would be nice to have the children do some role-play. So we told the boys that they were now to act like girls and the girls we told them to act like boys. The boys just loved it! They felt all powerful for they got to play the matriarch of the house… their grandmother! When we told them to go and play, the boys pretended to collect wood, fetch water, and cook and sow while the few older boys got to play the grandmother, watching and bossing over the rest at work. When we asked the girls to do the same, they just stood there and did nothing. They did not know how to play "boy." They said they felt useless. They got bored with the exercise. When the parents heard about this, especially the women, they were furious with me. They threatened that I could never come back again if I continued. We never did this exercise again.

Playing "boy" is just as much a challenge to girls as playing "girl" is a pleasure to boys. Chandan's "modern" woman of the village, the "modern" village if you will, can be argued to be alternative versions of a past. In a sense, consumption was not born in the modern era; it just gained a new form. Neat demarcations of the societal and cultural in modernity are confronted with ways of living that engage with and consume *Sony* and *Airtel* alongside *Kishore Kumar* Bollywood hits of the 1980s. Do we see paradox because of our discomfort as urban readers, fed on a staple of village norms that are bucolic and idyllic, primitive and romantic, overexposed to images of farm fields and herds making their way through sunsets by the barefoot villager? And what if this picture does come alive but for a little addition, a cellphone in hand, perhaps a watch on the wrist given by a cousin who has immigrated to Dubai to work in the oil fields, a T-shirt on a farmer that says Harvard? And what does "local" mean when Bihari immigrants run grocery shops and mend roads here, when woolens are imported from China at a cheaper rate than local handicrafts, and Sonia Gandhi is a household name? Can contradiction not be the definition of society and society an amalgamation of townships, and rur-towns? What does "Western" modernity do with swamis from Germany, British pilots and farming to digital versions of the Upanishads? In this work, expectations of modernity and technology for social change, of (over)interpretations of rurality and urbanity, of secularism and Westernization step aside while the story here walks by. And as it walks, we need to watch: its pauses, its turns; it bursting into a run, strolling, meandering, sauntering; of getting lost and found, of treading the road and at times getting off the path into winding alleys that are at once both laborious and pleasurable.

Chapter 4
New Technology, Old Practices

It's All in the Family

Suriola is a village about two hours off the road. Homes are scattered, with no straight route to get to the neighbor's house. Many of the homes are white washed, with painted red patterns at the entrance for good luck. Given that March to August is the season for marriages and festivals, these designs are ubiquitous. I am there with Champa to visit a *balwadi* gathering where the children are performing for their mothers (see Figure 4.1). It gives me an opportunity to talk to the village women afterwards. We are invited for tea to a woman's house. I am told she is of *thakur* (upper) caste. Outside the house, there are two cows, two goats and some chickens, evident signs of prosperity (see Figure 4.2). They own the property around them as well as a plot down the hill as part of their ancestral land. Cow dung has been patted dry and arranged in a pyramidal manner to be used as fuel for cooking. The eldest daughter-in-law is nursing her child outside. Her mother-in-law is playing hostess to Champa and me.

Figure 4.1   A *Balwadi* (Pre-School) Gathering

Figure 4.2   A Village Home

From outside, though neatly laid and large, the hut appears simple, made of stone and wood. Once you go inside, you encounter two large sofas, a color television, and a dining set with two cellphones being charged at the table. The television is on, showing a Hindi horror movie that is from the late 1980s. It is two in the afternoon. Nobody seems to be watching it. I am told that the eldest daughter-in-law doesn't do much housework as, "the poor thing, she has been living in the city for so long…she doesn't know the village ways anymore" (बेचारी, वह इतने लंबे समय से बाहर में रह रही है कि गांव के तौर तरीके नहीं जानती है). The younger daughter-in-law, however, comes from the neighboring village. She is expected to cook, clean, make the dung patties, and fetch firewood. This family also has a daughter who recently got married. However, her husband has tuberculosis and does not work. Fortunately, her family has some land and is able to give her money to fend for herself and her child. She is visiting. She brings in some tea. The television is still going on in the background, but this time running through a series of advertisements – of the latest *Lakme* whitening cream to make you "fair and lovely," *Airtel*, and *Vico* turmeric toothpaste, a promise that Ayurveda can still sell.

I ask how they brought all this furniture up through such narrow mountain paths. They say that they tied their furniture to the back of donkeys, including the television. "Yes, its difficult but at least we don't do it often…when people get sick, that's when we are in trouble…then we have to take them on a *doli* (cot-palanquin) all the way down to the nearest hospital on the road" (हां, यह कठिन है, लेकिन अच्छा है कि हम ऐसा अक्सर नहीं करना पड़ता। जब लोग बीमार पड़ते हैं तो मुसीबत होती है। तब हमें सबसे नज़दीक के अस्पताल पहुंचाने के लिए उनको पूरे रास्ते डोली पर नीचे सड़क तक ले जाना पड़ता है). Given this terrain and weather, particularly during the rainy season where there are frequent landslides, it is a common occurrence for women and children to slip and fall while on their way to the field. They sprain their ankle and sometimes even hurt their back. Then there is a need to carry them on a cot by two males, one on each end, to the nearest hospital. And what about making a phone call to the local doctor to have him visit, I ask. There are *jarpuks* (shamans) who they do call to help. Other doctors, they say, only stay in the town and won't even come to the hospital by the road, let alone their village. It is important to note that young men, given that they are a rarity in the villages due to the high migration rate, are valued particularly for their *doli* assistance. This family itself has no young men living at home. Both the sons are in the city, one in Delhi and the other in Chandigarh. This is a common phenomenon where men leave their wives and children in search of work and continue to send money back home. Even when they get work in the city, it could take years before the wife and children join them, if at all.

A boy, perhaps in his early teens, enters. He has just returned from school. His sister, perhaps a few years older, follows behind. They both are in uniform. Their school is a good few kilometers away, taking them almost two hours to get there. I ask them if their school has computers. Yes it does she says, but we hardly get to use them. There are only four computers; therefore, the teacher mainly gets the children to watch while she demonstrates Microsoft *Excel*, *Word*, and *PowerPoint*. They are also familiar with the *Encarta* encyclopedia CD provided by the government. They get about five minutes a day to use the computer:

We are supposed to get half an hour per week but then sometimes I never get my chance. Also, we do some small things on it like *Word* and all but that's it. The teachers don't know anything so how will they instruct? If you really want to learn computers, you have to take private lessons and pay 250–500 rupees for tuition and maybe even a certificate for a job.	हमें प्रति सप्ताह आधा घंटा मिलना चाहिये, लेकिन कभीदककभी मेरी बारी आती ही नहीं। इसके अलावा हम इस पर छुटपुट चीज़ें, जैसे वर्ड आदि करते हैं। लेकिन अध्यापकों को कुछ नहीं आता तो वे हमें कैसे सिखायेंगे? यदि वास्तव में कम्प्यूटर सीखना हो तो प्राइवेट संस्थानों में जाकर 250दक500 रुपये हर महीने सीखने के लिए देने पड़ेंगे और नौकरी के लिए प्रमाणपत्र भी हासिल करना होगा।

Their mother listens expressionless. I ask her what she thinks. She says she hears about computers but has never seen one, "what will I do with it at my age anyway? I barely know how to read and write, forget using a computer," (मैं अपनी इस उम्र में इसको लेकर क्या करूंगा? मुझे बहुत कम लिखना व पढना आता है, कम्प्यूटर चलाना भूल चुका हूँ।) she chuckles. She does proudly state that she talks on the cellphone with her husband in the city regularly.

Computers in rural schools are currently the rage. As has been stated earlier on, in the last ten years, multiple policies, both national and global have sought to address the issue of the digital-divide through the provision of computers to schools. But with that comes the much documented and predictable slew of problems, institutional and infrastructural: lack of regular electricity, dearth of trained teachers, poor technical maintenance, low standards, and inappropriate and inapplicable offline educational content (Arora, 2006c). Besides, prior to the intervention of computers, schools in rural areas have been critiqued for a host of issues such as teacher absenteeism, poor quality books, outdated educational content, distance between the school and the villages, gender discrimination and the like. These factors have contributed to the burgeoning of private schooling and the growing disengagement from government schools (Mehrotra, 2006).

Many government schools in Almora sadly live up to this notoriety. Take for instance the popular mid-day meal policy in schools that has received accolades and endorsements by the World Bank, many NGOs and the media. In essence, it has been well documented that poor children in rural areas do not have basic meals to sustain themselves. Thereby, providing lunch has been positively correlated with higher attendance and lower dropout rates (Drèze and Sen, 2002). As with all good intentions though, reality comes in the way. As a parent in a village remarks:

The government is supposed to give grains for the teachers to cook but very often you will see the grains going straight to the teacher's house or sometimes the government will send porridge for the children and the teacher has to pick it up from the shop in the nearby town but then who will go and collect it and then take it up the hill to the village? So often this porridge will be eaten away by insects or they feed it to the cows. Even if the food comes to the school, sometimes two hours go into the whole preparation process. And where my children go, two hours just goes into this food issue then why am I sending my children there? And then they make the food so badly and sometimes very dirtily. So many times my children come home with an upset stomach. So we say that we will make the food and give but you teach our children.

सरकार अध्यापकों को ग़ल्ला (बच्चों के लिए) पकाने को देती है, लेकिन अक्सर आप देखेंगे कि अनाज सीधा अध्यापक के घर को जा रहा है या कभी सरकार बच्चों के लिए दलिया भेजती है, जिसे अध्यापक को निकट के कस्बे की दुकान से उठाना पड़ता है लेकिन कौन इसे लेने जाए और पहाड़ की चढ़ाई में गांव तक लाए? इसलिए अक्सर दलिया कीड़े खा जाते हैं या वे गायों को खिला देते हैं। अगर खाना स्कूल तक आ भी जाता है तो कभीदककभीइसे पकाने के सारे काम में दो घंटे चले जाते हैं। ऐसे में मेरे बच्चों का क्या होगा? दो घंटे केवल खाने में निकल जाते हैं, तो मैं बच्चों को वहां किसलिए भेज रही हूँ? वे खाना बहुत खराब ढंग से और गंदगी से तैयार करते हैं। कई बार मेरे बच्चे घर वापस आते हैं तो उनका पेट खराब रहता है। इसलिये हम कहते हैं कि खाना हम बनाकर दे देंगे, लेकिन आप हमारे बच्चों को पढ़ाओ।

Teachers, in fact, are perhaps the most contentious issue of the lot, given the fact that much of what constitutes a "school," rests on the shoulders of the teacher.

Teaching jobs in government schools are one of the most sought after jobs in the village. As with many government jobs, they provide security, often regardless of performance. Another complaint surfaces:

There is this teacher you see. She is so bad and lazy, she does not even show up to class sometimes for weeks at a time. We got fed up and decided to complain to the education head office in Almora. We brought our complaints and requested to transfer her from there. The Education Department sent their people to check on her and realized that we were right so they agreed to transfer her somewhere else. But she had her brother-in-law sitting in some high position in Dehradun office who must have made some calls. Then the teacher said that she will not go…she said what will you do to me? What could we say? And what could the Department do? So she stayed.

इस ि िक्षिका को देखो। वह बहुत बुरी और आलसी है, कभी तो लगातार हफ्तों तक कक्षा में नहीं आती। मायूस होकर हमने अल्मोड़ा के जिला ि िक्षा कार्यालय में ि िकायत करने का नि िचय किया। हमने ि िकायत की और उसके स्थानान्तरण का अनुरोध किया। ि िक्षा विभाग ने अपने लोगों को उसकी जांच करने भेजा और पाया कि हम लोग सही थे, इसलिए उसे अन्यत्र स्थानान्तरण करने को राजी हो गये। लेकिन उसका जीजा किसी बड़े पद पर देहरादून आफिस में कार्यरत था, उसने जरू र कुछ टेलीफोन किये होंगे। तब ि िक्षिका कहने लगी कि वह नहीं जायेगी। उसने कहा, तुम मेरा क्या कर लोगे? हम क्या कहते? और क्या विभाग करता? इसलिए वह बनी रही।

While investigating computer usage, it is impossible to disassociate it from chronic schooling issues. It is tempting to demonize teachers, the government and development agencies that set the trend of policies and practices. Yet, there is a complex interplay of micro and macro policies and realities: desperate teachers trying to eke a living, fathers scared for the security of their girls when going to school, of teacher salaries subtracted for misuse or damage of books and computers, often resulting in these resources being locked up as a safety measure to other unintentional and often unpredictable outcomes. Yet computers are here to stay. Rather than continue to view usage of computers through the much exhausted lens of formal institutional failure, this study seeks to look outside of schools and into common public spaces of computer access and usage to best understand the range of learning that *does* go on with such tools. In the forthcoming chapters, the reader will witness a deliberate skirting of school issues unless and until it finds its way through discourse and practice with computers in a range of settings, situations, and spaces.

The Darling Child of Development: The Cellphone

In talking of computers and rural development, it is impossible to not address mobile phones. Another favorite technology for development, arguably the key favorite, cellphones have taken the world by storm. They have found their way into the hands of fishermen in Kerala as they check weather and market prices (Jensen, 2007), devout Muslims in Malaysia as they position themselves in the direction of Mecca with the aid of their mobile (Bugeja, 2004), and women weavers in Guatemala as they calculate their bank balance (Villereal, 2007). A Chinese family at a funeral ceremony no longer just burns paper items of the "basics" to ensure a good life for the dead in the next world but also burns paper versions of cellphones, laptop computers, and flat-panel televisions (Ikels, 2004). Mobiles are now part of the ceremony of social life. It overtook fixed lines in 2002 and its connectivity exponentially grew to 1.3 billion worldwide or by 80 percent in the last 4–5 years. In Almora itself, the penetration rate is about 85 percent, where most people in rural Almora have access to mobiles, if not own one.

What is fascinating about cellphones is that while computers are being consciously deployed by governments worldwide as a key strategy for rural development, mobile phones has been imbibed by people in rural areas with little governmental momentum. This local demand has made cellphones a more appealing choice as a means to reach the "marginalized." So, while both computers and cellphones are sought to initiate social change, there is a key difference: the former is supply driven while the latter, demand driven. While the private sector has been instrumental in coming up with creative schemes to extend their outreach to this new and exponentially growing consumer base, the government is by far, in the lead. The State has established much of the rural telephony infrastructure, laying grounds for private mobile providers to enter the fray. In fact, competition is so intense, including in Almora, that it has made mobile usage common place yet confounding in their maze of schemes. A shopkeeper that deals primarily with cellphones remarks:

> *Reliance* [private mobile company] and *Airtel* [private] can take off and leave but where will BSNL [government] go? They are the government so they will stay even in the villages. That's why BSNL is dominant but their service is very bad. You see their reception and call droppage is a lot but people bought it initially because they got a good deal but now it's regretful. Nowadays all these phone companies are getting to be the same…what's the difference between Tatacom, *Idea*, *Vodafone* and *Airtel* huh? Maybe the difference is Abhishek Bachchan [Bollywood actor] for *Idea* and SRK [another Bollywood actor] for *Airtel* [laughs]!

It is tempting to believe that the private sector is more efficient and thereby, able to gain the confidence of even rural users, but this is not necessarily so. For instance, I purchased a *Vodafone* phone card for half-price calls for 200 rupees, but after scratching the card and entering the pincode, it did not work. I went back to the shop I bought it from. It was run by three young boys, perhaps in their early twenties, all busy playing solitaire on the computer while listening to Bollywood songs from the CD on the computer. I (R) gave the card to the shopkeeper (S):

R: I want to return this card and get a new one.

आर. : मैं यह कार्ड वापस करके नया लेना चाहती हूं

S: Sorry madam, you've already scratched it.

एस. : माफ करना मैडम, आपने इसे खुरच दिया है

R: Well, I had to scratch it…how else would I know if it worked or not?

आर : मुझे तो इसको खुरचना ही था, नहीं तो मुझे कैसे पता चलता कि यह काम कर रहा है या नहीं?

S: Madam there is nothing we can do, it's nobody's fault, these things happen.

एस. : मैडम, हम कुछ नहीं कर सकते, इसमें किसी की ग़लती नहीं है, ऐसा होता है।

R: I paid for it so should I not receive the services? Why should I pay for something which is not my fault?

आर. : मैंने इसके लिए रुपये दिये हैं, इसलिए क्या मुझे सेवाएं नहीं मिलनी चाहियें? जिस चीज़ में मेरी कोई ग़लती नहीं है उसके लिए मैं क्यों रुपये चुकाऊं?

S: Madam this is how it goes.

एस. : मैडम, ऐसा ही होता है।

R: Why can't you complain to your contractors or sellers at least then?

आर. : आप अपने ठेकेदारों या विक्रेताओं से इसकी िकायत क्यों नहीं कर सकते?

S: What can they do madam, they are just middlemen. They themselves don't have any way of knowing. You have to complain directly to the main headquarters.

एस. : मैडम, वे क्या कर सकते हैं, वे तो केवल बिचौलिये हैं। खुद उनके पास ही कोई जानने का कोई उपाय नहीं है। आपको सीधे उनके मुख्यालय में िकायत करनी पड़ेगी।

And so the story goes. Given that the average income for a villager is 50 rupees (one dollar) a day, these systemic navigations can come at quite a price. In spite of some of these hurdles, mobile usage continues to grow substantively.

Cashing in on Technology

And if mobile usage amongst villagers does not sound futuristic enough, what about the new State Bank scheme to churn out biometric cards for rural banking. These "smart" cards have been promoted by the State Bank of India since 2000, yet are slow on the uptake. The manager though sells the idea:

> We recently put three ATM kiosks in Almora. The youth are taking to it like fish to water but the elderly are basically technophobes. Even my own father who is highly

educated has no idea of using an ATM card. We have also started these smart cards like biometric cards where all your information is there on it – we are in the process of establishing partnerships with shopkeepers in each of the small towns. All the villagers have to do is to put their thumb print and we put all the bank information and credit into that card and give him. Then when he wants money, he can go to the nearest shop from his village where a shopkeeper will have a computer. The villager will give him the smart card. The shopkeeper will double check to see if it's the same person with the thumb print as the photo will display. Then the villager can tell him how much he wants and the shopkeeper will dispense the money based on the account balance. Of course there is a limit. And this is good for the shopkeepers too – he will get a commission for each transaction and we will also get to save costs of building, hiring more people, salaries, pensions and all.

Mr Joshi, an elderly man, denies that he is averse to new technology:

I like going to the bank instead. You can talk to the bank manager and he will remember you. Why before people went to the bank, they did not mind the 15–20 minutes half hour wait…people could chat with people…it was a good time pass. And as if the lines have gone away, you just go see outside, there will be a long line for the ATM but no one will be talking to each other, just simply standing out on the crowded street instead of being inside.

इसके बजाय मुझे बैंक जाना अच्छा लगता है। आप बैंक मैनेजर से बात कर सकते हैं और वह आपको याद रखेगा। पहले लोग बैंक जाते थे तो 15दक20 मिनट या आधा घंटा इंतजार करने में कोई आपत्ति नहीं होती थी; लोग आपस में गप ाप करके अच्छी तरह समय बिता सकते थे। और मानो लाइनें खत्म हो गयीं, आप बाहर जाकर देखो, ए.टी.एम. के लिए लंबी लाइन होगी, लेकिन कोई भी आपस में बातचीत नहीं करेगा, बस एक भीड़ भरी गली में खड़े हैं।

While Deewan Singh states frankly:

Villagers don't have money and so they don't need to save or spend money. My father goes to the bank maybe once a year for his pension. And even if you have some money, the fees are still more at the end than what you deposit…why should we pay to have someone keep our money and charge us for it? It's better off under my mattress…there are people in my village who haven't even seen a jeep so what are we talking about ATMs?

ग्रामीणों के पास रुपये नहीं होते, इसलिए उनको रुपये बचाने या खर्च करने की जरू रत ही नहीं है। मेरे पिताजी भाायद सालभर में केवल एक बार पें ान के लिए बैंक जाते हैं। और अगर आपके पास कुछ रुपये हों भी तो जो जमा करते हैं उससे अधिक फीस होती हैं। इसलिए हमें अपने रुपयों को रखने के लिए किसी को भुगतान करने की क्या जरू रत है जो इसे संभालने के पैंसे ले? बेहतर है कि मैं इसे अपने गद्दे के नीचे रखूं। मेरे गांव में ऐसे लोग हैं जिन्होंने जीप तक नहीं देखी है, इसलिए हम ए.टी.एम. के बारे में क्या बात कर रहे हैं?

Perception is reality. It is worth remembering that new technology can and does get used for a continuation of old practices, just as much as old tools can gain a fresh perspective in usage as the situation changes. More importantly, what makes for new technology, including its features and affordances, can be discounted for the most part if the user chooses to do so. In other words, new technology does not necessitate new practice and may even complicate old practices. Regardless of all the features on the mobile, the wife in the village may just use it to call her husband once a week. She may even struggle with the mobile if someone were to modify the settings with the intent to ease her usage with this tool. Thereby, technology is as much a learnt practice as a social habit. As we will see in the proceeding chapters, close attention is paid to these relationships as it delves into the range of learning that goes on with computers.

Playing Low Key

Amidst the glitziness of new technologies lies a humbler tool of the trade – the plough. While some technologies "evolve" rapidly, some barely shift over decades, but not necessarily because they have reached their ideal form. In fact, there is no such thing as an "ideal" form but only that which is "sufficient" to the needs of those who get heard. This is about the politics of a technology: how much gets invested to solve a technological "problem" depends on who is being troubled, the political advantage they have in making their problem heard, their social status, how much can be invested to resolve that problem and more. We often find that when it comes to the relationship between rural development and technology, there is little emphasis on improving rural tools of the trade but rather, making urban the village setting with preference for developing tools that are seen as economically and socially mobile. Besides, old to new tools do not follow a predictable trajectory.

For instance, the G.P. Pant Institute, the main institute for rural development and agriculture in this area, came up with a machine plough some years ago for the farmers. To their surprise, they found tremendous resistance to this tool. The scientists were perplexed. One of them decided to spend some time in a village to understand this strange behavior. Through conversations with farmers, he realized that most of them felt ashamed to use it. They told him that the plough is pulled by an ox and now this new machine is pushed by a human, making them equivalent to those lowly beasts: "what will people say if they see me pushing a machine in the field," (मुझे खेत में एक म ीन चालते हुए देखकर लोग क्या कहेंगे?) one farmer said. Eventually, after much persuasion, the Institute was able to persuade one man, Krishnan Singh to adopt and try it out. And he did. Six months later, the scientists followed up on him. He was a happy man. He said that some people came to see his field and were impressed by how quickly he was able to get his field ploughed. So they decided to pay him to come to their field and operate the "machine." Other people in the village started to notice. Soon he was renting himself for 100–300 rupees depending on the size of the field. He got so many orders, he says, that he started to earn around 1,000 rupees (20 dollars) a day, a small fortune around here. By the end of the year, he sold his field and started to do this full time. He even bought many cows and started a dairy farm to sell milk, *ghee*, and butter. Today, he has a dairy farm and no longer uses the "machine." Here, the human became the "machine" for others and thereby, an acceptable piece of technology. Yet, resistance to direct usage of the mechanical plough endured.

Another piece of technology that continues to confound is the voting machine. In recent years, in fact, Almora, being mostly apolitical, has been caught in the web of partisan politics, of Congress versus the BJP. It is now widely known that, come election time, villages get deeply divided, sometimes making it impossible to do business or marital arrangements across villages that are politically opposed. Fortunately, politics is looked upon as a seasonal activity. With the political cycle over, people resume their neighborly ties.

The voting machine came here ten years ago and, prior to that, voting was done through the ballot fashion with paper and pen. This device runs on battery and not electricity. It has a chip on which registration and voting gets stored. It is periodically sent to Dehradun where the voting results are downloaded onto the computer. This allows for computational aggregation of the data into a larger feed of voting statistics which eventually circulates back to the villages through media.

Women are the local media's favorite target. They encourage women to take time off to vote; to show their loyalty and concern for their children's education through voting. But as we have seen, machines are not just tools of trade but extensions of intent, that which is often not shared. Far from being neutral, technology is a means of persuasion, seduction, and remembrances. As one village woman complains:

We were told to first press this one button on top and then one at the bottom. And so we did that. How were we to know that by doing that, the vote was already cast. This is how they tricked us. They made us vote for their party. The other time we were told that the machine is not working because of no electricity so we should go home. When we got back, my son told me that the machine does not even run on electricity!

हमें बताया गया था कि पहले सबसे ऊपर के इस बटन को दबाना है और फिर सबसे नीचे के इस बटन को। इसलिए हमने वैसा किया। हमें कैसे पता चलता कि ऐसा करने से हमारा वोट पहले ही पड़ चुका था। इस तरह उन्होंने हमारे साथ छल किया हमसे अपनी पार्टी के लिए वोट डलवा लिये। दूसरी बार हमसे कहा गया कि बिजली नहीं होने के कारण म ीन काम नहीं कर रही है इसलिए हमें घर लौट जाना चाहिये। जब हम वापस गये तब मेरे लड़के ने मुझे बताया कि म ीन बिजली से चलती ही नहीं है!

Eventually, we hope, as intent comes through, the learning that circulates can shift the position of the user to allow for the co-manufacturing of intent by the user herself. Perception of usage interacts with a multiplicity of users, spawning a wide range of outcomes that can be celebratory, confining and/or complimenting of ones day-to-day activities. In rural development, it's often not the market that dictates how the tool gets to be initiated. What is worth recording is that regardless of the outcome, the tool by itself cannot say much. It is its unique arrangement with other social actors and events within a larger matrix of beliefs and expectations that makes a compelling story. Rajesh, a social worker with 11 years of field experience remarks on rural schemes involving technology in Almora:

> IT [information technology] cannot just be placed in a community and expected to work. See, for instance, the government sees that we are in a disaster prone area so they say let's give the villagers money to build watershed areas that are 10x10 and spend 18 lakhs [36,000 dollars] in total. These are artificial water ponds for local irrigation and their target to complete this is by 31st March deadline of next year. Now what happens is that the government is very smart with IAS officers thinking big thoughts for the area but when we see Almora and the size of the landholdings, you will see that few are as big as 10x10 so where will they build these ponds? What they don't keep in mind is that the joint family of three brothers separates the land and each gets a holding, one 3x3, one 5x5 and one 2x2. So now what? Where will this pond fit? So the idea basically is good in theory. There is no water so the government should help these people. Scientists from Lucknow suggest 10x10 artificial ponds. This probably was feasible and worked in Punjab but the question is, is it suitable for Almora agriculture? But see, who has the time to talk to us or to any of the villagers when they design these plans? We are in the 5th zone, an earthquake prone zone. We also have landslides here. So what they also need to ask is that given this, is this large size necessary or can we make them smaller to suit the irrigational needs, the landholdings and our environment here. So basically, whether it is horticulture, agriculture or IT, development needs to become more of a broad science and think the big picture.

With hardware comes an emphasis on "software" – the people. There is a need to capture an ongoing dialectic between technology and the people to see what is being learnt, at what points and for what purposes. And *if* there is a need to produce a certain desired outcome, we need to first see what organically happens when the two intersect; differences in needs and positions of actors produce a range of outcomes. As Norman (1990) states, the outcome is less about the technical affordances that a tool comes with but about the actual social process it undergoes. This work asks the readers to remind themselves that computers are not just tools but a social phenomenon. They are continuations of past tools and approaches to these tools, resistance and embracement, perplexity and simplicity that circulates and surprises. Technological predictability is but a development myth.

Chullah and the Pump: Gender and Technology

There is a story behind the water pump and the *chullah* (stove), two common rural technologies with expectations of simple outcomes (see Figure 4.3). Both can be argued to be gendered technologies as they address specifically the woman's plight in rural villages. The former addresses the fact that women in villages, particularly Almora, have to walk a good length to reach their water source; the latter addresses the fact that women are inflicted with respiratory illnesses due to their burning of wood indoors for cooking.

To address this, the pump was installed a few years ago in one of the villages by an NGO. The hope was that women would be relieved to have a water source close to them where they could pump water at a predictable space. The next step was to install water pipes to their houses so they could get water directly without walking any distance. This led to diverse outcomes. The pump became a convenient source for the neighboring resort to exploit the water resource most effectively for their clients, capitalizing on this free and well-organized technology for water access and consumption. Also, there was a certain distain towards the pump, strangely enough coming from young women. It was found that they gained some respite from the watchful gaze of their mothers-in-law by taking long walks to fetch water. Also, given that most of these newly married women came from outside their

Figure 4.3   Village Water Pump
Source: Suresh Bisht

husband's village, the opportunity to fetch water with other women allowed for some bonding with their neighbors, whom they otherwise rarely see. This social isolation is partly due to the scattered settlements as well as the fact that, for most of the day, they are inundated with domestic chores and farming responsibilities.

Some women showed even less interest in having a water connection at their home. When taps were introduced into some of their homes, young women were deeply unhappy. When I suggested that perhaps they could still continue to meet their neighbors at a common space to socialize, they were aghast at the notion of meeting without purpose; "how can you just meet to talk? You need to do some work and then if you talk that's okay, otherwise you're just wasting time" (तुम सिर्फ बात करने के लिए कैसे मिल सकती हो? तुम्हें कुछ तो काम करना चाहिये और तब तुम बात करो तब तो ठीक है अन्यथा तुम सिर्फ वक्त बरबाद कर रही हो।). As for the *chullahs*, it is a basic technology that has persisted for many decades in spite of its evident ill-effects on the respiratory health of these women and often their children who are in the house while they cook. In response to this health crisis, the government, in partnership with a local NGO, came up with a policy to disseminate electric *chullahs* across some key village households as part of a pilot project. Villagers were given these *chullahs* but within sometime, these "smokeless" *chullahs* seemed to have left many of these "new clients" dissatisfied, resulting in reverting back to the "traditional" smoke *chullahs*. So what happened? Did these women not care about their health, some social workers wondered aloud. Did they not care for their children's health?

As with any technology, when embedded in a particular practice, there is a likelihood of it shifting and modifying other social practices that accompany it, for better or for worse. In this case, the advantage of the "traditional" smoke *chullah* was in warming the hut and thereby, being an efficient fuel utilizing device. Of course, this didn't work in all households as it was dependant on whether or not the chimneys were treated to prevent the heat from leaving the house. Besides, the traditional *chullah* is seen as "free" in the sense that firewood is "free" and available in the forests, with no need to pay for a gas connection. This applies if people subscribe to the view that women's labor of collecting firewood is not "work" in an economic sense, sadly a common perception (Shiva, 1991a).

The idea that the electric *chullah* is a "solution" to a problem is itself worthy of investigation. This idea, however "rational," is not often shared by all. Given that men don't work at the *chullahs* and that women rarely, if ever, express complaints about the *chullah* smoke, it gets perceived as less of a problem, "these *chullahs* have been used for generations and if they've [mothers, grandmothers, great-grandmothers etc] survived, so will my wife," (इन चूल्हों का कई पीढ़ियों से इस्तेमाल हो रहा है और यदि वे हमारी माताओं, दादियों व पर दादियों आदि। बच गये है तो मेरी पत्नी भी बचेगी।) exclaims the husband of one of the women who continues to use the traditional *chullah*. This links to the larger issue of preventive versus curative health, especially amongst women who seek for help primarily when it becomes an emergency. If the traditional *chullah* doesn't create an immediate problem, it makes for a weak case to substitute the traditional technology with a new one.

A simpler issue that contributes to the preference of the smoke *chullah* over the electric is the dependency factor. With electric *chullahs*, it's not just the cost of a gas connection; it is the dependency on other aspects for it to function. Since food preparation is at the heart of domestic chores, it is deemed too risky to leave it to chance. Dependency stems from the fact that the electric *chullah* can break or malfunction, in which case there is little one can do to fix it. In fact, dependency is a critical factor in technology usage across the board. What has to be figured out is what and who the tool is dependant on, for how long, when and within what contexts.

Ironically, some women argue that they are better off than their sisters or cousins or neighbors in the plains such as in Punjab where mechanized farming across large areas is common place. They argue that technology has supplanted much of their work, reducing the "usefulness" of the women in the plains. As one woman puts it:

My second cousin, she just cooks and takes care of the children. I do so much more out here. When she comes here, she can't even fit in because she has lost her skills, she doesn't even know how to work in the field properly anymore. What's the point? Her husband must think she is useless.	मेरी दूर की चचेरी बहन, वह सिर्फ खाना बनाती और बच्चों को संभालती है। मैं यहां उससे बहुत ज्यादा काम करती हूं। जब वह यहां आती है तो यहां फिट ही नहीं होती क्योंकि उसमें हुनर बचे ही नहीं हैं, उसे खेत में ठीक से काम करना तक नहीं आता। इससे क्या निकलता है? उसके पति को अव य ही सोचना चाहिये कि वह किसी काम की नहीं है।

Technology is tied to self-perception of usefulness and to the larger issue of social status. In fact, it is a double-edged sword: it threatens and pacifies, substitutes and extends, liberates, and at times, confines.

PART II

Computers and Rural Development

Chapter 5
Goodbye to the *Patwaris*

At least once a day in this village of 2,500 people, Ravi Sham Choudhry turns on the computer in his front room and logs in to the Web site of the Chicago Board of Trade. He has the dirt of a farmer under his fingernails and pecks slowly at the keys. But he knows what he wants: the prices for soybean commodity futures. (Waldman, 2004, *New York Times*)

Peasant Revolutions of the Past and Present

India is primarily an agrarian society in spite of its Silicon Valley reputation. The majority of Almora's population reside in 939 villages, where rurality and agriculture continue to go hand in hand (Guha, 2000). Its scattered and small land holdings across a difficult hilly terrain mark Almora as primarily subsistence-based. The average family in the village has about less than two hectares of land to cultivate, which doesn't allow for much irrigation (Agrawal, 2005). As a whole, the region has a low per capita income, which, combined with a high population growth, applies further pressure on using the land for survival.

It does not help that in recent decades, Almora, much like the rest of the Himalayan region, has experienced the siphoning of its rich resources to the neighboring urban areas (Trivedi, 1995). A historic low investment in infrastructure, industry and communication has exacerbated its desperate situation. It has been widely argued that these are the key factors contributing to peasant angst and mass movements that led to the formation of the new state of Uttarakhand. This new Statehood, after all, is meant to attract an influx of capital and enhance the hill region's position on negotiation for their resources.

Peasant movements and mass activism are hardly new to this area. Prior to independence, farmers rebelled against the British government's decision to appropriate their forest land for commercial purposes, leading to a successful and historic victory for farmers gaining communal rights over the forests (Farooqui, 1997). Other mass movements followed, some against feudal control and other more recent post-independent protests against what is seen as unjust State conversion of cultivable areas into forest land. Today, around 90 percent of the land has been taken by the forest department, leaving little for farmer cultivation. A pyrrhic victory, one might add; a boon to ecologists and environmentalists at the price of shrinking access to land for cattle grazing, fodder procurements to cultivation, damning farmers to their meager holdings.

Currently, these hilly regions are carved out for crops such as wheat, millet, barley, buckwheat, sugarcane, tea, oilseeds and potatoes, but it has not kept up with the area's population growth. In the area of sustainability, Almora seems to be facing an uphill battle. Explanations given for this are multiple. This area's moderate to heavy soil erosion has led to poor agricultural productivity (Bhatt, 1997). Popularly argued is that primitive agricultural techniques, combined with the lack of knowledge and appropriate technology, has exacerbated this situation (D.P. Agrawal, Shah, and Jamal, 2007). This is in alignment with the modernization perspective that has grasped policy makers post-independence to arguably the current times. There is a focus on converting "primitive mindsets" and rural "ignorance" through "right" knowledge for socio-economic mobility. This perspective clashes with the favoring of local empowerment that professes that inter-generational knowledge transfer on cropping has ensured a better nutritional balance and food security through variety and quantity. Women are seen as storekeepers of heritage and deep-seated knowledge of the land:

> This system was known as *baranaaja* where at least twelve different crops would grow on a single farm, and women were well aware of the different production practices including seed preparation of each variety of each crop. It led to better food security, especially as the region was somewhat isolated. Given the adverse geographical condition, it was considered best to grow as many varieties as possible in order to reduce vulnerability…more and more regions in Uttarakhand are becoming food insecure today with net-sown area, per capita food availability and access to food declining, especially in the hill districts. The availability of pulses (a major source of protein) and cereals (the most readily food available to the poor) has significantly declined…the key agents of change are the government and the market. (R. Agrawal, 2002, p. 7)

This argument is made against the State and corporations as they promote high-yield and hybrid varieties of seeds, cash crops, fertilizers and other inputs, without consideration of the long term consequences on the land and its people. Also, it is seen to negatively affect women's position in traditional agriculture.

Overall, much scholarship, primarily over the past two decades, highlights tensions and redundancies in the separation between "Western"/"standardized" knowledge and "indigenous"/"traditional" knowledge; the autonomy of modern knowledge against its traditional generational buildup. The scientific "expert" is pitted against backward rural know-how and, abstract policy-making against concrete and dynamic localized strategies of knowledge construction and implementation (Aikman, 1999; Chambers, 1983; Gill, 1996; Mgbeoji, 2006; Sillitoe, 2000).

New technology has a way of breathing life into old concerns. With computers seen as harbingers of the information age, new opportunities abound to correct old biases and wrong practices, and open up protected arenas. The State, perhaps tired of its domineering and "corrupt" reputation, promises the peasants a revolution from above, through massive public-private partnerships in disseminating new technologies to empower its rural citizens. Here, the State provides a solution to a problem of its own making – its infamous bureaucracy and middlemen. Corporations, both homegrown and abroad, vouch for new business models and more choices for the farmer where profit and social good go hand in hand. They espouse that circumventing middlemen by directly linking businesses with farmers will achieve lower prices of produce and higher pay for farmers. Meantime, NGOs strive to be part of this new momentum of the local supposedly reigning supreme.

New Intermediaries in the Making

Patwari in Hindi means a land record clerk. Their job entails visiting agricultural lands and maintaining records of ownership and tilling. They do not just keep the records of the land, they also measure the field area, evaluate its quality and in the event of a death and possible land dispute, are called to validate and mark the rightful heirs to the property (Stokes, 1989). Given the prime importance of land for the farmers, it does not take much to imagine the power the *patwaris* have on the farmers. For decades, these government officials have enjoyed authority on such matters and have been the main intermediaries interacting with farmers. This made room for tremendous corruption and exploitation of the farmers through an open system of bribes as "official rates" for such tasks (p. 215). In fact, this system has been unchanged since the British colonial rule where, as a means to keep track of the collection of land taxes, they periodically administered village land ownership.

A recent pilot project of computerizing agricultural land records is seen to have substantively reduced corruption in this area. In particular, the *Bhoomi* project in Karnataka, designed by the government's National Informatics Center (NIC), boasts of already having digitalized 10 million land records (Habibullah and Ahuja, 2005). While no doubt this process comes with its own challenges including access and usage of computers, it does provide an alternative for farmers, potentially freeing them from the stranglehold of the *patwari* system. This scheme gets the due credit for addressing long-standing and anachronistic ways of State information organization and processing. This initiative comes with visionary parallel initiatives, as has been noted in Chapter 2, of e-governance in areas of education, employment, taxes, and commerce. The computer as the new "middleman" is the new mantra.

In the name of village empowerment, much is being done. The State of Uttarakhand has decided to reinvent itself. It has launched a new IT portal named *Uttara*, a comprehensive information website on matters regarding government application forms, agricultural expert advice, agricultural market rates, filing of online grievances, weather reports, tenders, advertisements for employment, trade and general news. In particular, *Janadhar*, an initiative funded by UNDP in 2005, in partnership with IIT Roorkee, a top technology institute, aims at setting up community information centers (CICs) or *Soochna Kutirs* across its State. This is meant to serve as a one-stop-shop for information and services. As part of the e-governance drive, this programme has already been piloted in the neighboring Nainital district, with the closest center being just 45 minutes from Almora town. These centers are run on an entrepreneurship model, owned and operated by computer literate youth. Impact studies done by the NIC of the last two years of computer deployment, officially claim the following achievements: enhancement of farmer efficiency and productivity through IT, empowerment through information exchange amongst stakeholders, increase in demand for "agri-clinics," and a customer satisfaction index of 91.5 percent (NIC, 2005).

This pilot project is in the midst of being replicated in Almora. According to Om Prakash, the Secretary of Agriculture in Uttarakhand, the vision is to blend "domain knowledge with technology to enhance effectiveness and efficiency in agriculture and its allied sectors" (O. Prakash, 2007). Hence, the kinds of information that is made available through the *Uttara* portal for farmers include crop planning, husbandry, technology and process innovations, skills development, capacity building and access to virtual markets. More specifically, such an agricultural e-bank is meant to include detailed information on the following: geography of roads and location maps, agricultural department information, weather reports, *mandi* or government market rates, soil and seeds, crop life cycles, poultry, dairy and fishery techniques, horticulture and floriculture schemes, banking and insurance options, NGO details, new projects, and agricultural statistics. The government states that such an ICT project has been taken up to "provide relevant agricultural information in rural areas, helping farmers to improve their labor productivity, increase their yields, and realize a better price for their produce" (NIH, 2007).

Such commitment by Uttarakhand State has earned it the title of the "e-Merging State of India" by the Digital Empowerment Foundation in 2006. After all, Uttarakhand did not stop at the idea of computer kiosks. A Digital Library for Indian Farmers (DLIF) using open source software has been launched by the G.B Pant University of Agriculture and Technology. This scheme digitalizes texts on agriculture for farmers in multiple languages with an online audio-video component (Malhan and Rao, 2007). Other State-based initiatives include the World Bank funded project on "agri-clinics," and the *kisan* (farmer) call centers – a toll free service providing immediate agricultural information to farmers. There are also agricultural programs disseminated through radio, television, cable, magazines and journals. Doordarshan (National channel on television), the All India Radio (AIR) and the Indira Gandhi National Open University have for decades been sources of information to the farming community.

Global IT firms have entered the fray too, seeking to be at the forefront of perhaps the most ambitious computerization effort amongst the "emerging" markets. In the name of social entrepreneurship, corporate partnerships for digitalization of information take root. Microsoft's Digital Green (DG) is one such initiative targeted at farmers to provide agricultural information. Its unique selling point is its content repository that is primarily video-centric, argued to be more accessible to a large semi-literate and illiterate adult population. It is currently being populated with various types of information such as testimonials of farmers sharing their experiences, alternative income generating activity, experts leading step-by-step demonstrations on new agricultural methods, meteorological data, and marketing and government program information (Donner et al., 2008).

On the more home grown corporate side, perhaps the most written about and widely known e-agricultural initiative in India is the *e-chaupal* (Annamalai and Rao, 2003). ITC is one of India's leading private companies, with a market capitalization of nearly US 19 billion dollars, earning the reputation as "Asia's Fab 50." It is one of the world's most respected companies according to *Forbes* Magazine. ITC is also one of India's largest exporters of agricultural products. Its pioneering *e-chaupal* initiative has attracted much attention as it claims to revolutionize India's agricultural market by empowering Indian farmers through the power of the Net. This scheme intends to directly connect farmers with consumers and traders online through an exchange of information and commerce. Here, the computer operator is the sole intermediary. This scheme is meant to create transparency, and improve farmer productivity and income. ITC seeks to provide an alternative to *mandis*, the government market place for the farmer by offering better prices and more outlets.

The system typically works like this: farmers need to transport their produce to *mandis* or government-run markets, where government agents have fixed prices for the produce. The government agents then transport the produce to larger markets, particularly in neighboring cities, and sell them at a large premium to retail stores and small groceries. Through this system, the farmer has to bear the transportation costs of the produce. Given the long travel from the village to the market, farmers have little choice but to sell their produce once they arrive, often resulting in payments that are 10 to 15 times below market price; "...the middlemen are the ones earning a nice profit margin, while the farmers struggle" (IFMR, 2007).

The *e-chaupal* model works as follows: the entrepreneur/operator at the computer kiosk aggregates the village products and transmits the order to an ITC representative. During harvest time, ITC offers to buy the crop directly from the farmer at the previous day's closing price. The farmer then transports his crop to an ITC processing center, where the crop is weighed and assessed for quality. The farmer is then paid for the crop and given a transport fee. "Bonus points" are given for crops with quality above the norm. A farmer selling directly to ITC through *e-chaupal* can receive a higher price for their crops than through the *mandi* system. In this way, the *e-chaupal* system provides an alternative to government markets.

Overcoming computing access and usage barriers is definitely a challenge in such e-agricultural initiatives. However, even if this challenge is met, other assumptions pervade: that farmers will not contest information that they receive online; that they will trust these new computing intermediaries and old intermediaries will disappear. It is assumed that farmers will be more receptive to agricultural information when receiving it online versus through other traditional communication media such as the radio and television. Further, it is believed that the dearth of relevant agricultural information is the prime reason holding farmers back from achieving mobility. Another presupposition is that given the access to such information, farmers will abandon their "traditional" practices for "better" agricultural practices, where efficiency and productivity are the prime goals. To investigate such assumptions, it is important to first pay attention to what farmers truly want regardless of whether the information gets digitalized or not.

E-Agriculture Solutions Coming to Town

Scaling of a pilot project barely indicates success. A government project is its own beast. It replicates on accord of political momentum and is not necessarily result based. A local official who had worked at the start on the Nainital CICs admitted that it was not necessarily attracting people as they envisioned. However, some centers were doing decently but barely keeping afloat. Besides, given the routine transfer of officials in the Indian government service, the official confessed that this project would probably be put to the wayside, "it's about project ownership…once the government official in charge leaves, the drive is lost…it's all about leadership."

I was sent to a center nearby that served as a working example of the *Soochna Kutirs*, about 45 minutes from Almora town. Apparently it was not possible to establish connectivity at the heart of the villages, given Uttarakhand's topography. Hence, nearby towns were chosen based on their location to best serve a confluence of villages. Due to the limited government budget, the select locations are often not as strategic and desirable as the center of town, where most private cybercafés are stationed. Interestingly, besides government signage outside these centers, it would be hard to differentiate them from typical cybercafés. After months of visiting both cybercafés and CICs in this area, it is evident that the latter is far less frequented. In fact, neighboring cybercafé owners hardly see CICs as competition in spite of their supposed advantage of being subsidized and connected to the government; "yes I know that the government has set up some computer kiosks but they are no competition because no one goes there," a private cybercafé owner states, "…villagers come to us for it's a better choice for them." This is true to an extent as private cybercafés, being situated close to one another, brings about intense competition. This gets them to specialize and cover a wide spectrum of services such as email, printing, faxing, and typing forms and manuscripts, as will be investigated further in Part III section of this book.

Given that making a trip to town is expensive and time consuming, villagers prefer to come to a convenient location where they can fulfill multiple tasks including visiting the market, post-office, hospital, temple and other important institutions, located at the heart of town. Another reason for the preference of private cybercafés over CICs is that CIC maintenance often gets caught up in government bureaucracy while private cybercafés can repair their equipment faster. Also, given that private cybercafé owners are often the local computer experts, they are hired by the CICs themselves to repair and service their stations. Instead of this being viewed as a conflict of interest, cybercafé owners are grateful for an additional source of income. After all, with cybercafés growing in number in this area, it becomes a harder business to sustain.

On entering the CIC or *Soochna Kutir*, I encountered two girls waiting for their computer class to begin. They were both from the town. There was one man checking his Gmail. There were six computers in total, with its usual accomplices of printers and fax machines. A young man behind the desk asked if he could help. He was the entrepreneur of this center. After explaining that I was interested in the government's *Uttara* portal and was sent there by a local official, he began to demonstrate the portal. This required a User ID login to access government information and forms. Upon my asking why most people came there, he explained that they mostly used the Net for Gmail, Orkut and instant messaging. After looking at the search history, I noticed that the most frequent websites accessed in the recent past were Fotec private, State bank of India, State Board of Education Exam results and Google. Quickly, the young man pitched his key service – for 300 rupees per month, people could learn the Microsoft package and how to browse the Net.

He admitted that the *Uttara* portal could be useful if it was updated regularly and was interactive, "people would want to use these services and pay for them, but these sites are either not working or need updating…it's been three years overdue for updating of the content" (लोग इन सेवाओं का उपयोग करना पसंद करेंगे और उनके लिए रुपये भी खर्च करेंगे, परन्तु ये वेबसाइटें या तो खराब हैं या इन्हें अपडेट करने की जरूरत है। वर्तमान में तीन सालों से विशयंदकसामग्री अपडेट नहीं हुई है और ये इंटरेक्टिव भी नहीं हैं, अतः आप इनसे केवल फॉर्म का प्रिंटंदकआउट ले सकते हैं). At present, customers could print out forms but they still would have to go to the government office to complete the transaction. The only way the entrepreneur was able to stay afloat was by offering IT classes in the afternoons on a daily basis.

In my recent years of research of such kiosk initiatives, a plethora of difficulties have been encountered, including poor and outdated online content, weak connectivity, poor maintenance, inconvenient location and the like (Arora, 2005). This corroborates with recent studies regarding the dissemination of computer centers across rural areas (Rangaswamy and Toyoma, 2006). This should not necessarily alarm as the idea of a pilot is meant as an experiment; failure translates to lessons learnt for continued and improved modification. Hence, one cannot conclude much from this experience except that it will take much more commitment, institutional and otherwise to create village netizens. That said, what needs to be duly questioned is whether these centers are based on a valid premise. Being such a cost intensive initiative, we have to enquire whether the interests of the farmers, the core target population here, are being genuinely met. Instead of following the trajectory of institutional failings on which many have already embarked, this investigation focuses on what kinds of informational help the farmers seek, how they make decisions and what kinds of knowledge they already possess and share.

Kisan Sangattans

Extended week long conversations across eight months amongst farmers can be an eye-opening experience. Farmers from Almora and the neighboring districts of Nainital, Pithoragarh, Bageshwar and Champawat were brought together by Uttarakhand Seva Nidhi Paryavaran Shiksha Sansthan (USNPSS), a reputed NGO, to stimulate possible farmer collectives to jointly address matters that concern them (see Figure 5.1). This is a refreshing departure from the way most NGOs approach development, often going in with a specific intent, mission and knowledge of how to "help" people.

USNPSS has an interesting history. It is a public charitable trust founded in 1967 and consolidated as a non-profit in 1993. In 1987, it was appointed as a nodal agency by the Department of Education, Ministry of Human Resources Development and the Government of India, to undertake locale-specific environmental education programmes both in rural schools and villages in the hill districts of Uttar Pradesh, now Uttarakhand. Their mission is to:

> ...develop, through education and action, cohesive communities empowered to create rich, sustainable lives for themselves and future generations. Our long-term commitment is based on our belief that sustainability will become a way of thinking throughout the region.[1]

Their years of community service was acknowledged through the *Mahaveer* Award in 2003 for excellence in the sphere of Education and Medicine and recently its director, Dr Lalit Pande, won the *Padma Shri* for environment education, one of the country's highest civilian honors. Since 1987, USNPSS has supported (both financially and in capacity building) more than 200 Community Based Organizations (CBOs), spread all over the State of Uttarakhand. Some of them have grown into large organizations that work independently; others continue to work with them.

Their ties with the government are well known. The former governor of West Bengal and Punjab, the late Shri B.D. Pande, is Dr Lalit Pande's father and a *Padma Shri* winner himself. It is not a coincidence that the NGO's board has a stellar cast: the former Education Secretary of Uttarakhand, the former Chief Conservator of Forests and the former Vice Chancellor of Kumaon University. This organization was the brainchild of Ashish Da from the Mirtola Ashram (much written about in Chapter 3) who started working extensively with the local community for upgrading farming techniques. For this, he was awarded the *Padma Shri* in 1992 by the Indian government. This political clout has enabled the work of this organization to be embraced and institutionalized over these last few decades, a rare feat and dream for many NGOs.

---

1   See Uttarakhand Seva Nidhi Paryavaran Shiksha Sansthan (USNPSS) website: http://www.usnpss.org/.

These *kisan sangattans* or farmer meetings are a relatively new phenomenon for this organization. Looking to address the plight of farmers that daunts policy-makers and corporations alike, USNPSS chose an open-ended goal of discovering what farmers truly want in this region. This is executed through the sharing of skills and knowledge, addressing grievances and, most importantly, gauging and stimulating interest and motivation amongst diverse farmers to create groups of interest. This is no easy task as we will see in the section below, given that it's often tempting for NGOs to provide the "answers" and devise specific strategies to "solve" problems.

Besides, the very effort of organizing and coordinating such a vast and diverse group was tremendous and, therefore, these meetings were able to take place approximately once in two months. The day-to-day meeting schedule was as follows: farmer sessions started from eight in the morning and went on till late evening with breaks for meals and tea, ending with live music that carried through until late at night. Farmers camped at the NGO guest house for the week, many meeting each other for the first time. Awkward silence transformed to friendly banter by the end of these weekly sessions. On average, 30–50 farmers attended such groupings. Each district was guided by a local community organizer that the NGO supported. The meeting attendees were covered for their transportation costs, food, accommodation and other expenses. Thereby, I was able to experience three such week-long meetings over the course of fieldwork, interacting and living with about a total of 160 farmers from Almora and the nearby area.

This was, of course, no chance meeting. As has been stated earlier, I was stationed at this organization for most of my stay in Almora. Yet, this particular pursuit of the farmer meetings was good fortune given that this was taking place organically instead of the more ethnographically manufactured focus groups to elicit group understandings, perceptions and sharing of knowledge. For the most part I sat and listened and, at times, questioned. During meal times however, the enquiry continued but on a more informal basis.

It needs to be noted that the demographic was primarily male even though there were some women, particularly girls who had come with their village elders in the ruse to experience the "town." This by no means indicates a lack of women's participation as sometimes it was a negotiated turn-taking with their husbands (encouraged and often mediated by the NGO). Someone had to stay at home and take care of the elderly and the children. So, most women came separately for the *mahila* or women's *sangattans* which were facilitated by Dr Pande's wife, Anuradha, on issues ranging from elections, microfinance, loans, schooling, and healthcare. Unfortunately, this led to further role defining, with farming issues becoming more entrenched in the male domain while health and education viewed more as woman's issues. The NGO admitted to struggling in trying to bridge the two by encouraging young girls to attend farmers meetings, with the hope that the next generation would feel more confident in crossing these gender lines.

Figure 5.1   Uttarakhand Seva Nidhi

Consensus, Contention and Circulation of Conversation

Every meeting started with local *Pahadi* songs initiated by Renu, a staff member at the NGO. This was meant to break the ice and get people to open up. People were seated alongside the room on mattresses on the ground, encouraged to sit apart from their village members in hope of them mingling with others. Yet, for the most part, farmers chose to sit with their village group. A few members of the NGO staff were present, taking notes and interjecting when needed. Dr Lalit Pande presided over all these meetings along with his wife, Anuradha, who often would connect the discussion to women's issues at every possible opportunity.

The first day consisted of an introduction round where each person stated their name, their village and the reason for being there. Based on their interest, each person went into detail on their prime concern. This was followed by the local community organizers giving a summary of recent activities, events and village issues. On the first day, the framework for the following days was laid out. This was further discussed during meal time and then, in the forthcoming days, farmers were divided into groups based on common interests to tackle some of their concerns. This entailed much team work and presentations before the group. In the last day or two, "experts" were identified to share their success in a systematic manner. An agenda for the next meeting and follow-up strategies in their home villages were outlined. Visits were planned to the "expert" farmer's village for live demonstrations of their recommendations.

The introductions give a cross section of interests and insights: the girls stated that they wanted to teach, make jam, grow apples and potatoes; the young farmers expressed an interest in "opening the future" with new cash crops, and learning about fruits and vegetables. Most stated that their highest priority was to learn about marketing strategies. The older generation of farmers expressed interest on water issues and the future of their children; the community workers were interested in forming farmer collectives. Some had spent time in the city and were looking for a way back into the village again, "I used to be a driver in Lucknow but always wondered if I could come back and live in my village again;" "…after four years in Delhi, I decided I wanted to come back and do more…maybe fish farming and help my village." Others wanted to be prepared just in case other employment options did not pan out, "I applied to the army but if they reject me, I will do farming so I want to have a new plan just in case" (मै लखनऊ में एक परिचालक हुआ करता था। परन्तु हमे ा यह सोचा करता था कि क्या मैं फिर से वापिस जा पाऊँगा या अपने गांव में रह पाऊंगा।)…(4 वर्ष दिल्ली में रहने के बाद, मैंने यह नि चय किया कि मैं वापिस आना चाहता हूँ और कुछ करना चाहता हूँ कि जैसे मछलींदकपालन और अपनी गांव की मदद करना चाहता हूँ। अगर यदि और कुछ काम की सम्भावनायें नहीं बनती है। मैंने सेना में भी अर्जी दी है परन्तु यदि वे अर्जी निरस्त कर देते है तो मैं खेती करूंगा ताकि उस समय के लिए मेरे पास कोई दूसरी योजना भी हो।).

The top issue of concern across these meetings revolved around the commercialization of crops. Farmers wanted to make the shift from subsistence to profitability. Few had done so. Their key question was, "what to produce next?" Ideas flooded in: floriculture is big business, someone said; fish farming is the most practical, someone chimed; kiwis are the new fruit getting top prices, another remarked. Some heard that millets are in demand in the city now. Millets, known as the poor man's food, is now a health food in some cities. A subsistence crop becomes cash overnight. Agriculture is subject to trends. In fact, over these sessions, what is remarkable is that many such suggestions went in parallel with current recommendations and government schemes, as well as contemporary market trends. For instance, the Himalayas, given its unique climate and topography, have been known to be ideally suited for fruits and flower cultivation. Yet, only in recent years, with the growth of the urban middle class in India, there has been a strong domestic market and government push for such pursuits (Jodha, 2005). But that's only tip of the iceberg. Farmers argued about trade-offs that floriculture entailed – by supplanting crops with flowers, you trade problems of insects and high cost of pesticides with transport issues that is critical to the lifespan of flowers, its pricing and sale.

Also, while farmers are often accused of being risk-averse, shared experiences revealed a nuanced and complex reality of decision-making; "kiwi was a new word for even me two years ago," (2 वर्ष पहले मेरे लिए भी किवी एक नया भाब्द था।) confessed a young farmer as he explained his recent success in kiwi sales, "I didn't even know it was a fruit, forget how to find the plant and grow it" (मुझे यह तक पता नहीं था कि यह एक फल था, यह तो भूल जाओं की उस पौधे को कैसे ढूंढा जाये तथा उसे कैसे बढाया जाये।). He bought it from Delhi and with the help of a few Delhi traders, learnt what he had to do. Traders thereby, are not necessarily exploitative actors but can be possible and strategic knowledge intermediaries for farmers.

But lest we soften too quickly towards middlemen, a community organizer reminds us of the vicious chain of events that farmers find themselves interlocked in:

Farmers can't convert damaged fruits into pickles and jams. They need to be fresh so the farmer, after harvesting, can keep his peaches at home for maximum 5–6 days before it starts to rot. He has to send it immediately to the market in Haldwani. He knows that it only has a lifespan of 5–10 days depending on how he stores it. He can in this winter temperature make it last for maximum 10 days. As soon as it reaches Haldwani, it needs to be sent immediately overnight in cold storage to Delhi. So he will go through the middleman who negotiates for him in Haldwani and will handle all this for him. All he can do is send his produce and wait. There [the market], the produce gets negotiated and packed in cold storage and sent. The contractor and buyer strike a deal and the contractor gets his cut and leaves. The trader can easily report wrong prices to the farmer and the farmer has no choice and way of finding out the truth. Often, the contractor and trader delays payments for months. These farmers don't even know about interest rates on their own money. Instead, they borrow a loan from the buyers and pay interest on it while their money comes through which often take months.

किसानों के लिए सड़ चुके फलों का आचार और जैम बनाना कठिन है। इस काम के लिए फलों का ताजा होना आव यक है, ताकि किसान उन्हें अधिकाधिक 5दक6 दिनों के लिए ही अपने घर पर रखें, क्योंकि उसके प चात तो वे खराब होने लगते हैं। उसे उन्हें तुरन्त ही हल्द्वानी में बाजार में भेजना पड़ता है। किसान अपने हालात को देखकर मात्र 5दक10 दिनों के ही लिए फलों को सहेज सकता है, जिसके बाद वे तुरन्त ही हल्द्वानी पहुंचा दिये जाते हैं। हल्द्वानी पहुंचने के बाद रातो रात ही उन्हें दिल्ली के कोल्ड स्टोरेज में भेजना होता है। इस सब के लिए किसान को एक बिचौलिये की आव यकता होती हैं, जो कि न केवल इस पूरे कार्य को सम्पन्न करे, बल्कि इसके लिए हल्द्वानी में बीचंदकबचाव भी कर सके। किसान केवल अपने फल भेजने भर में ही सक्षम होता है। वहां बाजार में दुकानदार और खरीददार के बीच हुये समझौते में बिचौलिया अपना हिस्सा लेने के बाद विदा ले लेता है। इसके चलते बेचने वाला व्यक्ति बड़ी ही आसानी से मनमाना मूल्य किसान के समक्ष रख सकता है। किसानों के पास सच्चाई जांचने और परखने कोई तरीका नहीं रह जाता हैं। कई बार किसानों का पैसा महीनों तक उन्हें नहीं दिया जाता है, जिसकी वजह से उन्हें खरीददारों के ऋणी होना पड़ता है। अज्ञानताव । यही किसान सत्यंदकअसत्य से अंजान ही रह जाते हैं।

Thus, "ignorance" on price is hardly a matter of lack of knowledge. It's a lack of alternatives; a dearth of negotiating platforms for the farmer to stand on. The market often favors the interests of the middlemen. This is not to say that farmers don't continuously search for means to counter such mechanisms.

The very existence of these meetings is a case in point. It seems to be common sense that if these farmers unionized or became cooperatives, there would be far less exploitation. Farmers did see the merit in collectives, "we should all be on the same page. We all talk about coming together and working together but to do that, we should agree," (हम सब लोगों को एक साथ होना चाहिये। हम सब लोगों को एक साथ एक ही बात बोलनी चाहिये और एक साथ काम करना चाहिए लेकिन इसको करने के लिए हम लोगों को एक मत होना चाहिए) said a farmer from Champawat. But for every enthusiastic statement to collectivize, apprehension abounds. Dr Pande himself warned the group of such challenges:

Singing together for *holi* [Indian festival] is easy and fine but when money comes into the picture, there is no working together. It's easy to agree in discussions but when money comes in, that's when the problems also come. The big companies in the city do well because they have learnt to work together. Here, in our own town, we can't even make a line for the taxis or agree on a fixed price. Why is that so? Doesn't it make more sense for both the taxi drivers and the passengers? Here, people want to work together but they want instant results too and then there is a lot of infighting. So why not do what cities do? Fix prices and make a line of taxis. As the passengers come, they can get in line and pay the fixed price. The issue basically is trust. We should start with trust that is first and foremost internal and not external. You and you [pointing to two different farmers from two different villages] both know something but if you both farm individually, then both of you will suffer.

साथ मिलकर होली गाना आसान और ठीक है, लेकिन जब रुपयों का मामला आता है तब साथ मिलकर काम नहीं होता। चर्चाओं में सहमत होना आसान है, लेकिन जब रुपये आते हैं तब साथ में मुसीबतें भी आती हैं। भाहरों में बड़ी कंपनियां अच्छी तरह काम करती हैं, क्योंकि उन्होंने वहां साथ मिलकर काम करना सीखा है। हमारे ही भाहर में हम टैक्सियों के लिए एक व्यवस्था नहीं बना सकते या किराये की एक निर्धारित दर पर सहमत नहीं हो सकते ऐसा क्यों? क्या यह टैक्सी ड्राइवरों और यात्रियों, दोनों के लिए ज्यादा मतलब की बात नहीं है? यहां लोग मिलकर काम करना चाहते हैं, लेकिन उन्हें तत्काल नतीजे भी चाहिये और साथ ही बहुत सारे अंदरूनी झगड़े हैं। तो करते क्यों नहीं जो अन्य भाहर करते हैं? किराये की दरें तय कर दो और टैक्सियों को लाइन से भेजो; फिर जैसेंदकजैसे पैसेंजर आते हैं, वे लाइन में लगकर तय किराया दें और सबका नंबर आ जायेगा। मुख्य मुद्दा वि वास का है। हमें भाुरुआत भरोसे के साथ करनी चाहिये। वो सबसे पहली और जरूरी चीज़ है जो भीतर से आती है न कि बाहर से। आप दोनों को ही कुछ चीज़ें आती हैं, लेकिन यदि दोनों अलगंदकअलग खेती करने लगो तब दोनों को ही परे ानी होगी।

Here, consensus and collectivity are inseparable. For trust to take root, what is being asked of the villager is complete cooperation. For a "fairer" market system, villagers are asked not to compete but cooperate with one another. For instance, they can do so by withholding their goods for a better price, share their know-how, and jointly bargain with the traders. Voices of disagreement came forth, "…if it didn't work during my grandfathers time, why will this work now?" (अगर इसने मेरे दादाओं के जमाने में काम नहीं किया तो क्या अब यह काम करेगा?) quipped one. Another shared his village's struggle in forming a collective:

We were trying to form a union, one representative from each village but that organizer got sick so that went down under. Another time, we got some money together for group loans but that person managing our money betrayed us and ran away with it. We even tried to get the government to set a minimum price for our produce. Still, when we give them our produce, they busy eat away more money and do nothing else.	हम एक यूनियन बनाने की कोि ा ा कर रहे थे, हर गांव से एक प्रतिनिधि, लेकिन संगठनकर्ता बीमार पड़ गया और कोि ा ा थम गयी। दूसरी बार हमने समूह में कर्ज़ देने के लिये मिलकर कुछ रुपये जमा किये, लेकिन जो आदमी हमारे रुपयों का प्रबंध कर रहा था उसने हमें धोखा दिया और रुपये लेकर भाग गया। हमने यहां तक कोि ा ा की कि सरकार हमारे उत्पादन के लिए न्यूनतम मूल्य निर्धारित करे, लेकिन अब भी जब हम उन्हें अपनी उपज देते हैं तो वे और ज्यादा रुपये खाने में व्यस्त रहने के सिवाय कुछ नहीं करते।

While many farmers acknowledged the formidable challenges associated with forming a cooperative, some saw it as a necessary marketing strategy; less idealistic and more pragmatic. A young farmer from Nainital defended this idea:

To say and do is possible but to be successful is another matter. We need to try. These buyers we all know are not stupid. They advertise high rates for our produce and then when we get to the market, they say, look there is so much supply so we will reduce the price. We are then forced to sell our goods at low prices because we can't afford to transport everything back. But what if we were to stick together? What choice will they have then?	कहना और करना संभव है लेकिन सफल होना अलग बात है। लेकिन हमें कोि ाभा करने की जरूरत है। क्योंकि हम सब जानते हैं कि ये खरीददार मूर्ख नहीं हैं। वे हमारी उपज के लिए ऊंची कीमतें प्रचारित करते हैं और फिर जब हम बाजार पहुंचते हैं तो वे कहते हैं कि माल बहुत ज्यादा आ गया है इसलिये हम कम कीमत पर खरीदेंगे हम कम कीमत पर उपज बेचने को बाध्य होते हैं, क्योंकि हम अपनी सारी उपज वापस लेकर नहीं जा सकते। लेकिन यदि हम इकट्ठे हो जायें तो क्या ऐसा होगा? तब उनके पास क्या विकल्प रहेगा?

In response, a farmer described the "catch" in dealing with an NGO that was trying to encourage farmer's collectives by guaranteeing them fixed "fair" prices:

They say that they want us to get what we truly deserve and promise us higher prices than the mandi [government market]. But they only want the top quality produce	वे कहते हैं, हम चाहते हैं कि आप लोगों को वह मिलना चाहिये जिसके आप वास्तव में हकदार हैं। वे मंडी की अपेक्षा अधिक मूल्य देने का वायदा करते हैं, लेकिन उन्हें सिर्फ सबसे अच्छी गुणवत्ता की उपज चाहिये।

and they have fixed amount purchases per farmer as they want to buy a little from everyone. So instead of buying all the produce from one person, they buy only two of 20 quintals. Hardly ten people benefit from this scheme and that to barely. In the end, we have to go to the market anyway. When we get there, they [traders] say, what you have only the bad produce for us? Forget it, we will go with someone else. So then we waste money going to both places. It's not worth it.

उन्होंने प्रति कृषक खरीद की मात्रा निर्धारित की हुई है, क्योंकि वे हरेक से थोड़ांदकथोड़ा खरीदना चाहते हैं। इसलिए एक किसान से सारी उपज खरीदने की बजाय वे 20 कुंतल में से केवल 2 कुंतल ही खरीदते हैं। इस योजना से मुि कल से 10 लोगों को फायदा मिलता है और वह भी थोड़ा सा। अंत में हमें बाजार जाना पड़ता है, जहां हमसे कहा जाता है कि क्या तुम्हारे पास हमारे लिए सिर्फ खराब माल ही बचा? छोड़ो, हम किसी और से लेनदेन करेंगे। इस तरह दोनों ही जगह हमारा वक्त बर्बाद होता है, इसलिए ये बेकार हैं।

The challenge, thereby, resides in a system of incentives to keep quality high and yet be able to provide a fairer market system. So leveling of information and skills is the answer then? Some farmers see it's more complex than that. Not everyone can engage in the same thing after all, "it's impossible to do only one thing...we have to learn to diversify and get rid of the herd mentality" (यह सिर्फ एक काम करने के लिए असम्भव है...हमें भिन्नंदकभिन्न कार्य करना सीखना चाहिए और भेड़चाल की सोच को छोड़ना चाहिए।). Becoming a collective and producing the same thing are not necessarily two sides of the same coin. Production diversity can keep unions together. Strategic cooperation is needed for issues that concern them all – transport, state taxes, and dealing with middlemen. As one farmer suggested:

If a fisherman sells his fish quietly, then no one benefits. He has several challenges but he can't ask for help because he didn't share initially. Collectively though, the village can become specialized in something – like we can be known as the fish village and then as one group, we can benefit greatly.

यदि कोई मछुआरा चुपचाप अपनी मछली बेचे तो किसी को फायदा नहीं होता। उसके सामने बहुत सी चुनौतियां होती हैं, उसे औरों से मदद नहीं मिल सकती, क्योंकि उसने इसे साझे ढंग से भूरू नहीं किया। लेकिन सामूहिक रूप से करें तो गांव किसी खास चीज़ के लिए पहचान बना सकता है, जैसे हमें मत्स्य गांव के तौर पर पहचान मिल सकती है और फिर समूह में हमें काफी लाभ मिल सकते हैं।

Another farmer pointed out that effective marketing could happen better with a cooperative:

We can have the best produce but if we put it in our regular storage containers, few traders give us good prices. You see, those farmers who sell well put so much money

हमारा माल सबसे बढ़िया हो सकता है, परंतु इसे साधारण स्टोरेज कंटेनरों में रखने पर कम ही व्यापारी हमें बढ़िया दाम देते हैं। आप ही देखिये, जो किसान पैकेजिंग में ज्यादा रुपये लगाते हैं, वे जानते हैं कि

into packaging. They know that even though quite a lot of their money goes into the packaging, people will buy it because they will trust the product more. One man though cannot do everything, I can find prices, I can invest and I can package – is that possible? No, but together we can work as a cooperative to make a business I think.

चाहे इसमें बहुत रुपये खर्च हों, लेकिन लोग माल खरीदेंगे क्योंकि उन्हें इस पर ज्यादा भरोसा है। फिर भी एक आदमी सब कुछ नहीं कर सकता, मैं अकेला ही दाम मालूम करू , निवे ा करूं और पैंकेजिंग करूं क्या यह संभव है? नहीं, लेकिन मेरा सोचना है कि व्यापार करने के लिए हम मिलकर सहकारिता से काम कर सकते हैं।

To share new information, however, farmers have to gain new information. Constant learning is vital for survival. Some claimed that, "we need to see to learn…learning by watching is the only way" (हमें सीखने के लिए देखने की जरूरत है। देखकर सीखना ही एकमात्र रास्ता है।). Others professed getting outside their comfort zone for new ideas, "if we're willing to travel for our *netas* [politicians] from one place to another during their campaigns, why can't we travel more for ourselves, our associations?" (जब हम नेताओं के लिए उनके अभियानों के दौरान इतनी दौड़धूप करते हैं तब अपने आप के लिए, अपने संगठन के लिए इससे ज्यादा क्यों नहीं कर सकते?). In fact, farmers pressed on about new learning opportunities in agriculture:

We need to constantly learn, go to places and learn about things…in marketing, by traveling, we can sell better. We see what's possible. We also need to ask why. If we see an apple tree, we should ask if this can grow there, why not in my village – in this way, we learn.

हमें निरंतर, जगहंदकजगह जाकर चीज़ों के बारे में सीखना पड़ेगा। मार्केटिंग को लें, दौड़ध ूप करके हम बेहतर ढंग से बेच सकते हैं, हमें देखना है कि क्या हो सकता है। हमें 'क्यों' पूछने की भी जरूरत है; हम एक सेब का पेड़ देखते हैं तो हमें पूछना चाहिये कि अगर यह वहां हो सकता है तो मेरे गांव में क्यों नहीं? इस तरह से हम सीख सकते हैं।

But learning is not necessarily done for learning's sake. Most farmers saw new information as a means to make money:

In my father's generation, they were willing to struggle. But in my days, our children want more options. If you do farming, we can survive. But if you want to live, you need money. We need to do things differently now.

मेरे पिताजी की पीढ़ी के लोग संघर्ष को तैयार रहते थे, लेकिन अब मेरे बच्चों को अधिक विकल्प चाहिये। अगर हम खेती करें तो टिके रहेंगे, लेकिन यदि जीवन जीना है तो रुपये चाहिये। अब हमें भिन्न तरह से कामों को करने की जरूरत है।

Some new ideas require a change, not just in thinking, but in habits and lifestyle. This is a legitimate concern as families may not be willing to make such trade-offs. For instance, one of the community organizers happens to be successful in creating fisheries on his land. This endeavor capitalized on his land space as well as enriched his soil. He discovered that in shallow water, often accumulated in the hills, it was possible to breed small fish and sell them. However, there was much reservation towards this idea as fish was not part of the typical diet of hill people. Often, when families are unable to sell their produce, they consume it. This is the culture of subsistence farming, there cannot be any wastage.

Learning is also about asking questions, and paying attention to what's around you. That they did frequently and continuously, demonstrating a willingness to think of the impossible – of not being a farmer anymore:

Most of these [Almora] stores are started by Biharis. They're everywhere–tailoring, cementing, carpentry, construction, so why are we not starting these stores? If I want some masonry done, I have to beg these people to come and work. Why doesn't anyone want to work here anymore? Biharis are coming here and taking our jobs and we're leaving for elsewhere. Why should others come in when we're smart? We should learn what they can do. Our lifestyle is good here so why should we leave? If it's not agriculture, then something else.	अल्मोड़ा के ज्यादातर स्टोर बिहारियों द्वारा भुरू किये गये हैं; वे सब जगह हैं दक टेलरिंग, सीमेंट, बढ़ईगिरी, भवन निर्माण। हम इन कामों को क्यों नहीं भुरू करते? अगर मुझे चिनाई का कोई काम कराना है तो इनसे मिन्नत करनी पड़ती है। यहां का कोई व्यक्ति क्यों नहीं काम करना चाहता? बिहारी यहां आ रहे हैं और हमारे कामों को कर रहे हैं हम कहीं और को पलायन करके जा रहे हैं। जब हम करने में सक्षम हैं तो दूसरों को क्यों आना चाहिये? जो वे कर सकते हैं हमें उसे सीखना चाहिये; यहां हमारी जीवन भौली अच्छी है इसलिए हम यहां से क्यों पलायन करें? अगर खेती नहीं तो भायद और कुछ हो सकता है।

In fact, being a farmer is a mixed bag of emotion. Some viewed it as fatalistic, others as part of their heritage. Stories of struggle and unemployment pervaded these discussions: the tremendous odds against the farmer to change profession, inadequate education and the high trade-off of separation from the family through migration:

Farmer 1: We feel that if we get an education and don't get a job here, then we must leave, hoping to get a job in a company in Delhi. If we don't get that, we land up washing vessels in some hotel there. Isn't it better then that we open a tea stall here instead or work directly with the land we own?	किसान 1 : हमें यह लगता है कि अगर हमें पढ़ंदकलिखकर यहां कोई रोजगार नहीं मिल रहा, तब हमें दिल्ली किसी कंपनी में नौकरी की उम्मीद लेकर अव य जाना चाहिये और अगर वो न मिले ता वहां किसी होटल में बर्तन साफ करेंगे। क्या यह इससे बेहतर नहीं है कि हम यहीं चाय की दुकान खोल लें या अपनी ज़मीन में काम करें?
Farmer 2: The problem is that everyone wants to leave. There is too much pressure in this culture. It's about respect. If you're a	किसान 2 : समस्या यह है कि हर कोई जाना चाहता है। यहां बहुत ज्यादा सांस्कृतिक दबाव है। यह सम्मान के बारे में है। यदि आप खेती करते हैं तो अपने ही लोग आपकी इज्ज़त

farmer here, you don't get treated well by your own people. If you're in Delhi, and even if you're washing dishes, at least you're free from the taunting.

नहीं करते। अगर दिल्ली में रहकर बर्तन साफ करने का काम भी करो तो कम से कम लोगों के ताने सुनने से बचे रहते हो।

Anuradha: I agree. You've seen for yourself at home. If you have two daughters-in law, one helps at home and does everything and the other is in Delhi…when the Delhi bahu [daughter-in-law] comes to visit [the village], you treat her so nicely. You say, don't do this and don't do that, she is not used to it, and you give her so much respect and the one at home, you disregard. Do you treat the two as same? No!

अनुराधा : मैं सहमत हूं। आपने अपने ही घर में देखा होगा। अगर दो बहुएं हैं, एक घर पर सारा कामकाज करती है और दूसरी दिल्ली में रहती है; जब दिल्ली वाली बहू गांव आती है तो उससे इतना अच्छा बर्ताव करते हैं। कुछ काम भी नहीं कराते कि बेचारी को इसे करने की आदत नहीं है। उसे इतना अधिक सम्मान दिया जाता है, जबकि घर वाली बहू को इज़्ज़त नहीं मिलती। दोनों से एक जैसा बर्ताव किया जाता है? नहीं!

Farmer 1: That maybe true. The mistake people make is that they think they can leave for work outside and one day come back and buy land. By then, there will be no land to buy. Once you leave, you leave.

किसान 1 : हो सकता है यह सही हो। लोग यह सोच कर ग़लती करते रहे हैं कि नौकरी करने बाहर चले जाते हैं और एक दिन वापस लौटकर जमीन खरीद लेंगे, लेकिन तब खरीदने को जमीन बचेगी ही नहीं। एक बार आप गये तो फिर हमे ा के लिए गये।

Farmer 3: They [migrants] realize too late that working for others doesn't contribute to anything, health, life. If you leave the field, you will just wander around so better to farm than do a job in the city.

किसान 3 : पलायन करके जाने वालों को बहुत देर से समझ आती है कि दूसरों के लिए काम करने से वे स्वास्थ्य, जीवन इत्यादि में कुछ योगदान नहीं कर सकते। अगर खेतों को छोड़ देंगे तो खाली इधरंदकउधर भटकते रहेंगे। इसलिए भाहर में नौकरी करने से बेहतर है कि गांव में खेती करें।

The topic of migration seemed to tug at the heart strings. A mixture of fear, helplessness, resentment and envy form a complex reality. While most farmers criticized migrants, almost every family in a village in Uttarakhand has some member who has left home in search of a job. Also, when issues regarding women came up during these meetings, farmers for the most part agreed readily so as to not confront or delve further into discussion. Given that it is widely known that women are the backbone of agriculture in this region, there is still a strong gender demarcation concerning monetary aspects. As agriculture becomes commercial, it enters the male domain. Thereby, males see themselves as intermediaries of knowledge, guiding women in matters of agricultural production even when not directly working on the field.

Some farmers have sought a way out of farming to other employment opportunities, but with little success. Education is seen as a possible entry into a new career path. Yet, some question the usefulness of formal education:

Farmer 1: I've studied but what use has it been for me? All that knowledge and no job? Same with all this knowledge on vegetables. Say I learn all this but I have little control over my own land and rain, the kinds of seeds available, so then what's the point?

किसान 1 : मैंने पढ़ाई की है, लेकिन मेरे लिए इसका क्या उपयोग रहा? तमाम जानकारी है, लेकिन कोई नौकरी नहीं? ऐसा ही सब्जियों की जानकारी का है; कहो तो मुझे सब जानकारी है, लेकिन अपनी जमीन, उपलब्ध बीजों और वर्षा के ऊपर मेरा थोड़ा सा ही नियंत्रण है। ऐसे में क्या कर सकते हैं?

Farmer 2: Education does not mean degrees – it's about sharing advice. My hope is that if you start a coaching center in the village, that would be more valuable than any books. Books are not the only source. In India, it's become a habit to get degrees. Just for the certificate, you waste so much money but you learn nothing. Just a useless piece of paper.

किसान 2 : ि क्षा का मतलब डिग्रियां नहीं, जानकारी के आदानंदकप्रदान से है। मुझे यह उम्मीद है कि यदि गांव में एक कोचिंग सेंटर भारू किया जाय तो वह किताबों से अधिक मूल्यवान होगा। किताबें ही एकमात्र स्रोत नहीं हैं। भारत में डिग्रियां हासिल करना आदत सी बन चुकी है। सिर्फ सर्टीफिकेट के लिए आप इतने रुपये बर्बाद करते हैं, लेकिन कुछ नहीं सीखते। यह सिर्फ कागज़ का बेकार टुकड़ा है।

Farmer 3: These days these children, all they want is cricket, cellphones and cars. And how will they get that? That's all they are learning is what we can't have.

किसान 3 : आजकल के सारे बच्चों को क्रिकेट, मोबाइल और कार चाहिये। उनको ये सब कैसे मिलेगा? वे उन सबको सीख रहे हैं जो हमारे पास नहीं हो सकतीं।

Farmer 2: That's why we need to share our ideas in our own way. You have to know what you're talking about when you do something new. You decide on a new crop and a big scientist comes to you, then what? He will tell you things and you will not know what to say, so you will do what you are told. We need to know what we're doing.

किसान 2 : तभी तो हमें अपने विचारों को अपने तरीकों से आपस में साझा करने की जरूरत है। जब आप कुछ नया कर रहे हों तब आपको उसके बारे में जानना होगा। आप नयी फसल उगाने का फैसला करते हैं और एक बड़ा वैज्ञानिक आता है, फिर क्या? वह आपको चीजों के बारे में बतायेगा तुम समझोगे नहीं, फिर भी उसे सुनेंगे। जो काम हम करते हैं उनके बारे में हमें जानकारी होनी चाहिये।

Anuradha: You should then educate your girls too. They work in the

अनुराधा : हमें अपनी लड़कियों को भी ि क्षित बनाना चाहिये। वे खेत में काम भी

field. Let them in and bring them with you for these meetings.

करें, लेकिन उन्हें इन बैठकों में साथ लेकर आना चाहिये।

Farmer 3: If educating a girl helps me to get dal-roti [food] then I will even sell my house, otherwise, what is the point? I will have to spend money on her marriage and then she will go away and then? And besides, who will marry her if she gets too educated?

किसान 3 : अगर लड़की को पढ़ाने से मुझे दालंदकरोटी पाने में मदद मिले तो मैं उसके लिए मकान तक बेच दूंगा, नहीं तो क्या फायदा? मुझे उसकी भाादी में रुपये खर्च करने पड़ेंगे और फिर वह चली जायेगी। तब? इसके अलावा अगर वह बहुत अधिक पढ़ंदकलिख लेगी तो उससे कौन भाादी करेगा?

Computer education, in particular, is the new trend, the new disillusion. Such feelings are rarely about the artifact; it's more about its place in their lifestyle, the range of livelihood opportunities at a farmer's disposal, and the potential to create genuine change:

(Elder) Farmer 1: There is a polytechnic college of the government, an engineering college here where you also learn computers and other degrees but it's not for us. These colleges are for outside people, not us locals. The government spends crores [millions of dollars] in our name on these engineering colleges and says it's for us. But then, they charge 5 lakhs [10,000 dollars] a year. You also have the costs of books, hostel, and food. So in this area, among 400–500 students, only there are five students taking this degree, that too, mostly from Delhi. Basically, these computers are for the bank manager's sons and not for our children – yes, we do need to learn computers but who will teach us?

(प्रौढ़) किसान 1 : वहां एक सरकारी पॉलीटेक्नीक कॉलेज है, इंजीनियरिंग कॉलेज है जहां कम्प्यूटर व अन्य डिग्रियों के लिए पढ़ाई होती है, लेकिन यह हमारे लिए नहीं है। ये कॉलेज बाहरी लोगों के लिए हैं, हम स्थानीय लोगों के लिए नहीं। सरकार करोड़ों रुपये हमारे नाम पर इन इंजीनियरिंग कॉलेजों पर खर्च करती है और कहती है कि यह हमारे लिए है, लेकिन एक साल के 5 लाख रुपये लेते हैं। किताबों, हॉस्टल, और भोजन का खर्च भी हुआ। इसलिए डिग्री ले रहे 400दक500 छात्रों में से इस क्षेत्र के केवल 5 बच्चे हैं, वे भी अधिकतर दिल्ली के हैं। मुख्य रूप से ये कम्प्यूटर बैंक मैनेजरों के बच्चों के लिए हैं, हमारे बच्चों के लिये नहीं। हां, हम कम्प्यूटर सीखना चाहते हैं, लेकिन हमको कौन सिखायेगा?

(Young) Farmer 1: I spent 500 rupees to learn computers. After learning computers, the people who studied are still at home while the man who was teaching got a job in the factory and sold his computers.

(युवा) किसान 1 : मैंने कम्प्यूटर सीखने के लिए 500 रुपये खर्च किये। कम्प्यूटर सीखने के बाद भी लोग घर पर ही हैं, जबकि सिखाने वाले को फैक्टरी में नौकरी मिल गयी और उसने कम्प्यूटर बेच दिये।

(Elder) Farmer 1: In Delhi, each house has a computer but they wear masks on their face. We have no computers but we roam free! Can a computer give milk? What can computers give?

(Elder) Farmer 2: No, that's not a fair comparison.

(Elder) Farmer 3: I got this US flyer and it says that our land is being used as a dumping ground for computers. Already our land is small and our people are many and then we have to share it with the garbage of computers? What is the need for that?

(Young) Farmer 2: That way what's the point of knowing agricultural prices or anything about our produce these days when we have little to sell? We have to face the facts. Our landholdings are dividing up and we have little left so I prefer to do computers where I know there is a future.

Dr Pande: So the discussion here is why farming? Why computers?

(Elder) Farmer 1: Doing work with our hands, forest, fresh air, day-to-day self sufficiency is what will give us happiness. There is no relationship between computers and farming.

(Elder) Farmer 4: It's not about happiness but what we can do. Not everyone can get a job. There

(प्रौढ़) किसान 1 : दिल्ली में हर घर में कम्प्यूटर है, लेकिन वे अपने चेहरे पर मास्क पहनते हैं। हमारे पास कम्प्यूटर नहीं है लेकिन हम स्वतंत्र घूमते हैं! क्या कम्प्यूटर दूध दे सकता है? कम्प्यूटर क्या दे सकता है?

(प्रौढ़) किसान 2 : नहीं, यह सही तर्क नहीं है, यह उचित तुलना नहीं है।

(प्रौढ़) किसान 3 : मुझे यह विदे ी एन.जी. ओ. की पुस्तिका मिली इसमें लिखा है कि हमारी भूमि का इस्तेमाल कम्प्यूटरों के डंपिंग ग्राउंड के तौर पर हो रहा है। पहले से हमारी जमीन छोटी है और लोग बहुत सारे हैं, अब इसमें कम्प्यूटरो का कचरा भी रहेगा? इसकी जरूरत क्या है?

(युवा) किसान 2 : इस तरह जब हमारे पास बेचने को थोड़ा सा है तब कृशीय मूल्य या उपज के बारे में कुछ भी जानने का क्या औचित्य है? हमें सच्चाइयों का सामना करना है। हमारी जोतें बंटी हुई हैं और हमारे पास थोड़ा बचता है इसलिए मैं कम्प्यूटर को तरजीह देता हूं जिसमें मुझे भविश्य दिखता है।

डा. पाण्डे : तो यहां बहस इस पर है कि गाय का दूध कि प्लास्टिक दूध (सीधे गाय से प्राप्त दूध या व्यापारिक रूप से पैंकेट में बिक रहा दूध)? खेतीबाड़ी क्यों? कम्प्यूटर क्यों?

(प्रौढ़) किसान 1 : अपने हाथों से जंगल में काम कने, खुली हवा, रोजमर्रा की आत्मंदकनिर्भरता से ही हमें खु ी मिल सकती है। कम्प्यूटर और खेतीबाड़ी के बीच कोई सम्बन्ध नहीं है।

(प्रौढ़) किसान 4 : बात खु ी की नहीं बल्कि इसकी हो रही है कि हम क्या कर सकते हैं। हरेक नौकरी नहीं पा सकता। पुलिस में 2 पदों

are 500 applicants for just two posts in the police so most people have little choice but farming. Computers versus farming? Even if there is no link, it is good to learn computers. With computers you can save a copy of important paperwork. When it rains, our homes in the villages leak and sometimes even break, and our house articles get wet and the land records, our bills, and our hill certificates get ruined. This is our security. If we look at a broader level, if it's about our lifestyle and good life, then the link is there.

के लिए 500 लोगों ने आवेदन भेजे हैं, अतः ज्यादातर लोगों के लिए खेती के कम विकल्प बचे हैं। एक ओर कम्प्यूटर और दूसरी ओर खेती? यद्यपि इन दोनों में कोई संबंध नहीं तब भी कम्प्यूटर सीखना अच्छा है। कम्प्यूटर द्वारा महत्वपूर्ण कागजातों को बचा सकते हैं, क्योंकि जब बारि ा होती है गांव के हमारे घरों में पानी रिसता है, कभीदककभी घर टूट जाते हैं और भूमि संबंधी कागजात, बिल, सर्टीफिकेट भीग कर बर्बाद हो जाते हैं। यह हमारी सुरक्षा है। अगर हम इसे व्यापक स्तर पर देखें, अगर हमारी जीवन भौली और अच्छे जीवन की बात करें तो दोनों के बीच संबंध है।

(Elder) Farmer 2: Computers can be useful to buy goods but can't help us in farming.

(प्रौढ़) किसान 2 : कम्प्यूटर माल खरीदने में उपयोगी हो सकते हैं, परंतु हमें खेती करने में मदद नहीं कर सकते।

Dr Pande: Why?

डा. पाण्डे : क्यों?

(Elder) Farmer 2: Because we don't know about computers and there is little possibility that we will get to know.

(प्रौढ़) किसान 2 : क्योंकि हमें कम्प्यूटरों के बारे में जानकारी नहीं है और इसकी कम ही संभावना है कि हम जान पायेंगे।

(Elder) Farmer 4: There is a relationship and can be a relationship between the two. Computers can be useful in getting information on seeds, prices, water and cropping but who will go there to get it is another question.

(प्रौढ़) किसान 4 : इनमें संबंध है और दोनों के बीच संबंध हो सकता है। कम्प्यूटर से बीजों, कीमतों, पानी और फसल उगाने के बारे में जानकारी पाने में सहायता मिल सकती है, लेकिन इसके लिए वहां कौन जायेगा यह सवाल दूसरा है।

(Young) Farmer 2: A cellphone call for 1 rupee can tell me the same that a 20,000 rupee device does. Even then, tell me who calls to get farm information?

(युवा) किसान 2 : एक रुपये की टेलीफोन कॉल से मुझे वही जानकारी मिल सकती है जो 20,000 रुपये का उपकरण करता। तब भी बताओ कि खेती की जानकारी के लिए कौन फोन करता है?

For the most part, computers are not looked upon as extensions and intermediaries for new agricultural practices and knowledge. Instead, they are viewed as symbols of a new life, a break from an agrarian lifestyle, and a privilege of the urban class; their ability to open pathways to new employment, new skills, and for an alternative future for the farmer's children.

Also, the link between computers and agricultural information is made but given little weight. When there is emphasis on this relationship, it's usually seen through the lens of "expertise." Computers are seen as extensions of experts that tell the villagers what to do next. Farmers associate this with past "advice" that has led them astray, information that, when applied, did not work to their advantage:

Farmer 5: By listening to GP Polytechnic [local government research center on agriculture and technology], we can be in big trouble but we listen anyway. Like the other time, they gave us seeds for planting beans. It should have behaved like the previous crops but all the insects came into the field and ruined it. And what about the other time? Their seeds sprouted 15 days earlier, then the next year the seeds sprouted 15 days later and the third time nothing happened at all. Our first way of doing things, was it not useful? Was it not good? This comes from traditional practices but every time we leave [for the city] and come back, something is lost. So we listen to these people [at the university]. Is it then better than computer knowledge?

किसान 5 : अगर हम जी.पी. पॉलीटेक्नीक की सुनें तो बड़ी मुसीबत में फंस सकते हैं, परंतु फिर भी सुनते हैं। इस बार भी उन्होंने हमें सोयाबीन लगाने को बीज दिये। यह पहले की फसल जैसी ही होती, लेकिन तमाम कीट खेत में आए और इसे बर्बाद कर दिया; और दूसरी बार? उनके बीज 15 दिन पहले ही अंकुरित हो गये, फिर अगले साल बीज 15 दिन देरी से अंकुरित हुवे और तीसरी बार कुछ हुआ ही नहीं। चीजों को करने का हमारा पहले का तरीका क्या उपयोगी नहीं था? क्या वह अच्छा नहीं था? यह पारंपरिक व्यवहारों से निकलता है परंतु हर बार हम इसे छोड़ते हैं और फिर वापस आते हैं, कुछ खो गया है। इसलिए हम इन लोगों को सुनते हैं। क्या यह कम्प्यूटर के ज्ञान से बेहतर है?

Farmer 6: And what about the seeds for cauliflower? Last time their seeds gave us huge cauliflowers but what's the point? The families here are only 4–5 so we prefer our small local cauliflower so that way no wastage happens so why do we buy these seeds? It doesn't even taste as good. Because we keep thinking that we can sell this to the city but we don't. We have such little to sell that no real market is interested in our produce anyway.

किसान 6 : फूलगोभी के बीजों के बारे में क्या हो रहा है? पिछली बार उन बीजों से बड़ी फूलगोभियां हुईं लेकिन इससे क्या करें? यहां सिर्फ 4दक5 परिवार हैं इसलिए हम अपनी छोटे आकार की फूलगोभी को ज्यादा पसंद करते हैं जिससे कोई बर्बादी नहीं होती, इसलिए हम वे बीज क्यों खरीदें? इसका स्वाद भी उतना अच्छा नहीं है। हम यह सोचते जरूर हैं कि इसे भाहर में बेचेंगे, लेकिन ऐसा करते नहीं हैं। हमारे पास बेचने को इतना थोड़ा होता है कि हमारी उपज को खरीदने में किसी की रुचि ही नहीं है।

Farmer 7: Where is the security also? If they [government] decide to shut down their programme or when they give us hybrid seeds that don't sprout when it rains, even if it's free, we have lost a whole cropping season and we will be ruined anyway. Do you think they will compensate for this? If these private companies come to us to become the only buyer and shut down tomorrow, our investments are gone and then what's next?

किसान 7 : सुरक्षा भी कहां है? अगर वे अपना कार्यक्रम बंद करने का फैसला कर लें या जब वे हमें संकर बीज देते हैं जो बारि ा होने पर अंकुरित नहीं होता, चाहे मुफ्त सही लेकिन हमारी मौसम की पूरी फसल बर्बाद हो जायेगी और हम भी बर्बाद हो जायेंगे। क्या आपको लगता है कि वे इसका मुआवजा देंगे? यदि ये निजी कंपनियां अकेले खरीददार के तौर पर हमारे यहां आयें और कल को कारोबार बंद करके चली जायें तब हमारा तो सारा निवे ा बर्बाद हो जायेगा, फिर इसके बाद क्या होगा?

Farmer 6: We don't even get enough water, be it with our own seeds or hybrid so what is the difference? At least our local seeds, 50 percent will sprout as it's used to such hard conditions even with the little water but hybrid seeds won't. There are all sorts of government schemes now such as water catchments but only half fills up because there's no water. Now if I build it, I will only be able to use it next year so not many people like that. They want water now!

किसान 6 : हमें पर्याप्त पानी ही नहीं मिलता, ऐसे में हमारे अपने बीज हों या संकर बीज, क्या अंतर है? पानी कम भी हो तो हमारे अपने स्थानीय बीजों में से कम से कम 50 प्रति ात तो अंकुरित होंगे, क्योंकि ये ऐसी कठिन परिस्थितियों के आदी हैं, परंतु संकर बीज नहीं होंगे। आज कई तरह की सरकारी योजनायें हैं, जैसे जलागम, पर इनमें से आधी ही पूरी होती हैं क्योंकि पानी है ही नहीं यदि मैं आज योजना बनाऊं तो इसको अगले साल उपयोग कर पाऊंगा इसलिए अधिकतर लोग उसे पसंद नहीं करते। उनको पानी अभी चाहिये!

Farmer 7: Everyone wants everything NOW! We should start an ATM for water out here [many laugh].

किसान 7 : हरेक चाहता है कि सबकुछ अभी मिले! हमें पानी के लिए भी ए.टी.एम. भुरू कर देना चाहिये (बहुत से लोग हंसते हैं)।

Learning to Decide

The "information highway" is more a side street in the process of decision-making. Information plays a small part in the larger process of learning to decide. Middlemen, far from being the blockage to "essential" and "relevant" information, are a by-product of larger and often insidious systems at work. For instance, the price of soybeans is only one of the myriad concerns as farmers learn (and teach others) their limits in negotiations through the systematic encountering of price fixing between buyers and traders. Their ingenuity is found in accepting such constraints and working with them through bribery, and pioneering new marketing strategies to entice traders with their produce. At times, as we have seen, some farmers harness the strategic relationship between the buyer and trader by allowing the trader to dictate new production avenues, as in the case of the kiwi farmer, converting a possible adversary into a colluder in the making.

Hence, exploitation is less about scarcity of information and more about scarcity of choice. Choice is not just about more market opportunities but of furthering access to institutions and people that enable such choices. Farmer's knowledge encompasses the awareness of such scarcities. It becomes a matter of trade-offs. One farmer's choice of fish farming goes alongside possibilities of new domestic food consumption patterns. Another farmer's bias to join a cooperative takes on the risk of betrayal by members of the cooperative. An embrace of hybrid seeds can come at the high cost of crop failure.

Farmers reveal their reservations on collectives as group "agency" comes at a price. Diversity and specialization may mitigate some of these concerns yet, new issues are bound to crop up as such collectives require tremendous effort to be sustained. The NGO, as an intermediary here, can serve as an agent of knowledge. They allow for a common and temporal space of sharing of ideas, introducing new ways of thinking and, at times, reinforcing ongoing perceptions and beliefs. The agenda of NGOs can constrain, but in this case, has succeeded through an open-ended forum, in revealing a difference of approaches and beliefs. *Sangattan's* for *sangattan's* sake is an alien concept to many village folk. The romanticism of development through participation is hardly the inspiration behind farmer cooperatives. Instead, this is viewed as strategic marketing and specialization through the leveraging of common interests to form a stronger foothold for price negotiation. Such rationalization carries through in negating education for education's sake, most powerfully communicated in a farmer's logic of the negative investment in girl's education. Decision-making is about constructing a rationale. Information is subservient to perception and belief. Narratives of failed hybrid seeds, the "miracle" effect of chemical fertilizers, and the inaccessibility of higher education for farmers, contributes to the shaping of the local fact.

Small farmers are pressured to be more adaptable to new ideas and processes to survive. Intergenerational transfer of knowledge should not be looked upon as automatic. Particularly, in an area such as Uttarakhand where migration is a common phenomena, such disjunctures allows for "urban" notions to seep through. Rural thinking does not have the prerogative over "tradition." Farmers don't have "farmer" thoughts; men don't have "male" thoughts. Yet, the epistemology of the local versus global, and tradition versus modern persists in policy construction. As we have seen, farmers in Almora do not just deal with "agricultural" knowledge but with education, banking, markets, gender, trust, and community. Common talk of *Biharis* "stealing" their jobs, flyers on computer e-wastage, and the unmarriageable prospects of an educated woman become part of a complex knowledge system. Decisions are made just as much from fear and insecurity as through logic. In fact, one can argue that insecurity is the best rationale for deciding against innovation, against new processes and technologies.

Thereby, computers, the new middlemen, are subject to similar restraints and resistances as other intermediaries. After all, computers don't make decisions, people do. Granted, computer's unique appeal comes from its symbolization of a new way of life, a possible escape from rurality and a pathway to new employment. Farmers barely stated their thirst for new information from computers but for new computer skills for new jobs – computer related jobs. To profess that computers can revolutionize the field of agriculture is to negate decades of research on the slow rate of computer adoption by farmers in Western nations (Ascough, Hoagb, Frasierb, and McMastera, 1999; Doyle et al., 2003; Putler and Zilberman, 1988). This is not to suggest that farmers are luddites or that the "Western" ways are the norm. Rather, it is to point out that this blanket assumption that computers *will* be embraced *if* only populated with the "right" content for farmers disregards long-standing resistances as irrelevant. Instead, let's take the humbler view of computers as one of the multiple intermediaries such as the State, NGOs, traders, farmers, and children, competing to be the prime filters of information to enable decision-making.

Chapter 6

Excavating Relics of an Educational Idea:
The Romance of Free Learning[1]

Ethnographer as Archeologist

On a narrow lane just off the main street of Almora town, looking past the typical motley industry of tea stalls, fruit vendors and Kumauni woolens, is a cemented four-walled structure with three gaping holes staring back at you (see Figure 6.1). Its utter unremarkability could easily be mistaken as one of the many construction projects in Almora that ran out of money and, thereby, left with careless abandon. If you are not seeking for it, you could just pass this by, readjusting your lens to more pleasant sites of the Himalayan landscape.

Figure 6.1   HiWEL Learning Station in Almora Town

1   A version of this material has been published as an article in the *British Journal of Educational Technology*. I would like to thank Colin Latchem, the Asia-Pacific Corresponding Editor of *BJET* for his tireless effort in shaping this article. Also, I

Figure 6.2   HiWEL Learning Station in Hawalbagh Village

Instead, here, the ethnographer has become the archeologist, following rumors of relics of a technology initiative for learning that was a precursor to other IT projects in this region. Almost immediately on my arrival, I was directed to this abandoned computer kiosk site (with some satisfaction one might add, to some of the locals who believe this is typical of "doing good" in the name of technology). To them, this serves as living proof of the pointlessness of technology initiatives by the government, NGOs and the private sector.

These ruins provide a segue to another similar but more recent demise at a nearby village called Hawalbagh, where three kiosks stand in sleeping stupor, eyes shut to the world (see Figure 6.2). Both spectacles share the grounds of government intercolleges, the former right on the path to the boy's intercollege (high school) while the latter is situated on the school's playground.

Examination of these concrete skeletons leads directly to the famous Hole-in-the-Wall (HiWEL) experiment started in 1999 by NIIT, a private IT company, catching the international development world's attention and triggering romance with the idea of *free* learning – free from restrictions of formal learning – the

would like to thank David Hawkridge for his insightful comments and the anonymous reviewers for their feedback. I am much indebted to Sugata Mitra for challenging and engaging me in this novel and important arena of informal learning.

regimentation, the control, the systematic constraints exercised upon the child in her discovery through self-learning (Mitra, 2003). It started with a simple idea. A computer was sunk into a wall near an NIIT office in Kalkaji, New Delhi, to see what children would do with the computer with no instruction or guidance. The computer screen was visible from the other side of the wall. A touchpad was built into the wall. Windows NT operating system was installed on this PC. A video camera was placed nearby, on a tree, to record the children's activity. Children that used this kiosk came from the nearby slum with little formal education and fewer still with exposure to quality instruction, the English language and computers.

Within eight months, "the children of the neighborhood had learned all the mouse operations, could open and close programs, surf the Net and download games, music and video. When asked, they said they had taught themselves" (Mitra, 2004, p. 11). Mitra, the pioneer behind this brainchild, defines this as Minimally Invasive Education (MIE), a new pedagogic method "that uses the learning environment to generate an adequate level of motivation to induce learning in groups of children, with minimal, or no, intervention by a teacher" (Mitra et al., 2005, p. 2).

Here, HiWEL strives to provide digital access through computer-equipped learning kiosks in school playgrounds and out-of-school settings in the underlying belief that children can take ownership of their learning, and that learning can be driven by their natural curiosity. It is posited that this can pave the way for a new education paradigm and be the key to providing literacy and basic education and bridging the digital divide in remote and disadvantaged regions. It is also suggested that the child can both be the learner and the teacher, collaborators in their own learning. Not surprisingly, over the years, this idea has found its voice and funding through numerous national and international efforts, media coverage and publications (although, admittedly, by NIIT researchers for the most part), spawning ingenious means and ways of marking villages with their signature "playground computer kiosks."

As a joint venture between NIIT and the International Finance Corporation, HiWEL has received the World Information Technology and Services Alliance (WITSA) "Digital Opportunity Award" for its work in informal elearning. Nicholas Negroponte of MIT likened the Hole-in-the-Wall kiosks to "shared blackboards" which children in underprivileged communities could collectively own and access to explore, learn, collaborate, brainstorm, come up with exciting ideas and express themselves. Even the novel "Q and A" on which the *Slumdog Millionaire* movie was based was inspired by the HiWEL initiative. The author, Vikas Swarup, says "My book is about hope, optimism and triumph of the human spirit. I was inspired by the Hole-in-the-Wall project...That got me fascinated and I realized that there's an innate ability in everyone to do something extraordinary, provided they are given an opportunity" (*Economic Times*, 2009).

If this isn't good grounding for romance, then stories of children inventing their own vocabulary to understand computer symbols like the hourglass as *damru* (Shiva's drum), or the fact that kiosks are specifically designed with enough legroom for only children (Mitra, 2000) should evoke a sense of vindication for children, who often are relegated to the background even (or especially) in matters of schooling. In fact, this romance becomes full blown passion as we leaf through evidence of children flocking to such kiosks, teaching themselves and others to paint, play games and music and check out their horoscopes. In this voyage through computers, they discover learning *as* fun. So the question remains – *what went wrong in Almora?*

Digging Up the Past

Contact with the HiWEL staff confirms the diagnosis of these two "Learning Stations." The Almora kiosk, inaugurated on the 5th of October 2002, faced an untimely death within a few months of opening due to vandalism, stated as being the only learning station to be closed even before the project period was over by a HiWEL member. On the other hand, the Hawalbagh kiosk was less sordid in a sense that, having been initiated during the same time period, it had survived a few years and became inactive in 2007. Meantime, a "caretaker" in the village had been appointed to look after the equipment while the kiosk lay dormant. I was provided with the necessary contact details to help "revive" the Learning Station. I had been volunteered to assist HiWEL as I inquired into the status of its kiosks. It seemed that the New Delhi team was unaware of the reasons for the Hawalbagh standstill and awaited enlightenment on that issue.

Hawalbagh itself is a small village with a population of about 600, with literacy rates reported at 64 percent (Sati and Sati, 2000). Rampant unemployment exists here as there is little industry but for the agricultural research center, which employs a third of the population on an intermittent basis. Thereby, as with most of Almora, subsistence agriculture supports the majority of the population. There is only one primary school, one government intercollege and one private school in this area. Typical of the government schools in Almora region, Kumauni is spoken widely but Hindi is the medium of instruction and English is taught from 6th grade onward.

HiWEL's computer kiosk is located within the government co-educational intercollege compound, where more than 50 percent of the students come from neighboring villages. Talking to the teachers, there seemed to be a general lack of curiosity or concern for the fate of the kiosk. As mere observers, they remarked that it was in use for sometime and then stopped and now laid there, pointing to the silent concrete that stood on their playground. They recollected that a few boys would use these kiosks but, "usually for things like games, that's all" (ज्यादातर उन चीजों के लिए जैसे खेलों बस और कुछ नहीं।). Interacting with the children resulted in little insight too. They responded either with a lack of awareness of the kiosk or a nod in acknowledgment of its existence, as if accepting its fate as part of the general fate of things; a shrug of the shoulder, a brief comment of "just played around" (बस यू ही) with the kiosk or for the most part, a general ignorance of its past and presence. The hole in the wall was now in the ground, deeply buried and barely seen.

The "caretaker," on the other hand, was more informative. He happened to be the vendor for HiWEL. He was not paid to support or maintain the equipment. A young man in his twenties, he recollected clearer than the others of HiWEL, even though he hadn't heard from them in months:

The students mainly came from the inter-college and sometimes nearby colleges. At that time, it was going well...like 4–5 years...students were using it because their colleges did not have any computers and it was free. They [HiWEL] put a full time instructor there at the site to guide students. These students did a lot on their own together but once in awhile they would ask the instructor for help. But when the electricity bill did not get paid for awhile, the authorities cut it off. I told them [HiWEL] to pay but they did not get back to me. So I just put the computers in a safe room and locked it for now. They are still there.

छात्र मुख्य रूप से इण्टर कॉलेज और कभींदककभी आसपास के कॉलेजों से आते थे। उस समय तक, पिछले 4दक5 सालों की तरह यह अच्छा चल रहा था छात्र इसका उपयोग कर रहे थे, क्योंकि उनके कॉलेज में कम्प्यूटर नहीं थे और यह मुफ्त था। छात्रों के मार्गद िन के लिए संगठन ने स्थल पर एक पूर्णकालिक प्रि क्षक रखा। बच्चों ने अपने आप साथ मिलकर बहुत ज्यादा किया, लेकिन कभींदककभी वे प्रि क्षक से मदद मांगते थे। लेकिन जब कुछ समय तक बिजली के बिल का भुगतान जमा नहीं हुआ तो प्राधिकारी ने कनेक िन काट दिया। मैंने संगठन से बिल का भुगतान करने को कहा, लेकिन उन्होंने कोई जवाब नहीं दिया। अतः मैंने कम्प्यूटरों को एक सुरक्षित कमरे में रख दिया और आज तक इसमें ताला लगा है। कम्प्यूटर अभी भी वहीं हैं।

On sharing this with HiWEL, it yielded a response of their needing to "empower the community" to take care of the learning station and that there was an expectation that the *sarpanch* (the head of the village) would take care of this learning station but failed to do so. Three hundred rupees (about eight dollars) per month was all that was needed to sustain this learning station. They saw themselves as doing the "handholding" for the first three years and beyond which, they expected the "community" to take the baton.

Almora town yielded quite a different feedback. Given the short life of the kiosk due to the vandalism in 2002, few people around there including the students, had any recollection of this project. Walking down the winding path that led to the college in mid afternoon, I saw a group of teachers sitting around in the playground reading the newspaper and chatting. It was exam time and they had come out for a break. After carefully listening to my enquiries, one of the senior teachers took the lead to answer the question on what went wrong with the kiosk:

It was a good idea but am not sure why it doesn't work...I'll tell you, they spend so much money on computers and such little on taking care of it. They gave the

यह अच्छा विचार था, लेकिन मैं निि चत तौर पर नहीं कह सकता कि यह कारगर क्यों नहीं है, मैं आपको बताऊंगा कि उन्होंने कम्प्यूटरों पर इतना ज्यादा रुपये खर्च किये और इनके रखरखाव पर इतना कम। उन्होंने

keys to the chaukidar [watchman] and told him to clean it and take care of it. Now you tell me why should he take care of it? He doesn't get paid for it at all. He sleeps here below and works below at the school so why should he go up and stay up just for this?

चौकीदार को चाभियां सौंप दीं और उससे इसे साफ करने और इसकी देखभाल करने को कह दिया। अब आप मुझे बताइये कि वह इसकी देखभाल क्यों करे? उसे इसके लिए कोई वेतन नहीं मिलता। वह यहां नीचे सोता है और नीचे स्कूल में काम करता है, इसलिए वह ऊपर क्यों जाये और सिर्फ इसके लिए वहां क्यों ठहरे?

Another teacher piped in:

The problem is that there is no instruction given. It's okay if people are computer literate but when most people here are computer illiterate they need guidance and instruction. In our school we have a full lab where we instruct children. This kiosk thing went on only for a month or two and then it stopped but even though it stopped working we kept getting the bills for months and then only recently they took the computers away and moved it to Hawalbagh for something.

समस्या यह है कि कोई प्रशिक्षण नहीं दिया गया है। यदि लोग कम्प्यूटर साक्षर हैं तब तो ठीक है लेकिन जब ज्यादातर लोग कम्प्यूटर निरक्षर हैं तो उन्हें मार्गदर्शन और प्रशिक्षण की जरूरत होती है। हमारे स्कूल में एक पूरी प्रयोगशाला है जहां हम बच्चों को सिखाते हैं। यह किऑस्क सिर्फ एकदकदो महीने चला और फिर बंद हो गया लेकिन यद्यपि इसने काम करना बंद कर दिया हमें महीनों तक बिल आते रहे और अभी हाल ही में वे कम्प्यूटरों को उठा ले गये और किसी कारण से इन्हें हटाकर हवालबाग ले गये।

And so the story opens up. It is easy to get wrapped up in the usual ping-pong of development politics from community participation to corporate responsibility. However, this would not help much in addressing why the romance turned sour, a romance which *should* have gone right, *should* have had a happy ending. After all, who doesn't want to see children take over the driver seat in their own learning for a change? To best answer this query, we need to pay attention to the pillars of HiWEL for *free* learning: informal public learning, unsupervised access, collaborative peer teaching and self-guided learning. The key is to examine *free* learning and its relationship with formal educational institutions as well as the perception of computers, the spaces learning inhabits, and its social processes, enactments and performances.

School As You Go

The concept of free learning is not simply concerned with liberation from long-standing inequitable access to education. It entails the transformative capacity of learning that is more dialogic and less didactic (Freire, 1998). It disregards hierarchies and formal structures and promotes the alluring proposition that learning can take place anywhere and with anyone. It does not take much stretch of the imagination to draw linkages between such advocacies and the HiWEL experiment in trying to provide education without dependency upon the teacher and the school.

After all, the classroom can be seen as a suffocating as well as a nurturing environment. Through the school, the State attempts to achieve consensus on the voice of wisdom and learning deemed necessary for the socialization of a new generation. A school is therefore as much a conceptual as a concrete creation. The organization of learning through such a single agency can be seen as a political act. Foucault's "school as prison" analogy has become a well known emblem of the forces at play in creating learning spaces to assemble and shape human thought and action (Foucault, 1977). The four walls promise to close in at any time and this promise of fear, it is supposed, can drive schooling far more powerfully than any alternative vision. Thereby, school is perceived to be as much an instrument of political will as an embodiment of a vision of democracy.

So it is not surprising that HiWEL's free and open learning should excite many. It holds promise of a "minimal interventionist" grassroots approach to education with maximum benefit. Learning escapes the confines of the school walls, is available anywhere and anytime, and overcomes the years of injustice in educational access in countries such as India where the provisions for and expectations of so many children are low (Arora, 2006a). In Almora itself, stories abound of teachers handing over the keys to the classroom to senior students while they stay at home or undertake other work.

Of course, we are attracted to the promise that children can learn and do learn with no or little supervision using computers in environments free from the chronic barriers to achieving schooling in disadvantaged areas. But in the case of Almora and Hawalbagh, what we see is the idea of free learning going into free fall. In other words, the act of learning without conventional schooling constraints is contingent on the support of institutional, social and other factors, making it less "free" in that sense. Would the fate of the kiosks have been different if there had been more supervision? And would this have meant less or more freedom in learning? The vandalism of the kiosks (Mitra, 2004), in this case at Hawalbagh and their neglect at Almora suggest that too much freedom may be a bad thing. This is why HiWEL, while aiming for independence from the contemporary drawbacks of rural schools, is still compelled to choose sites on the school compounds, hereby associating the free offer with schooling, classes, and teachers.

Let's be clear about this. HiWEL does not aim to dismiss or disregard schooling; it seeks to provide an alternative space for learning in places where school systems have failed to provide for children. So there are two aspects to being "free" in learning: free from the dearth of educational support due to teacher absenteeism by giving children opportunities to learn with the computer kiosks and free from "bad" teaching by providing spaces for collaborative, self-organized and peer teaching and learning – making the child the agent in his/her learning process. In this formula, however, lies a fundamental paradox. HiWEL exists because there are few teachers in rural areas to guide students. HiWEL also exists because they offer an innovative Freirean pedagogy that opposes the rote linear learning that often takes place within rural schools. So at once, it is supportive of existing schooling practices and yet, against the way they currently stand, providing seemingly better pedagogic strategies to learning. So why would this duality be a problem?

The conundrum HiWEL has to face is that it has to strategically engage with schools to justify its presence due to the absence of instruction, and yet has to strategically disengage in the presence of certain types of instruction. So for HiWEL to sustain itself, it has to involve the school. However, if it involves the school to a point of carrying over certain rote and linear schooling practices, it is in danger of becoming nothing but an extension of that school.

This is seen, for instance, where HiWEL benchmarks its achievements in computer literacy, English language and other academic areas against conventional schools (Inamdar and Kulkarni, 2007). Thus while pedagogical expectations can be perceived here as being "invasive" on child-centered growth, HiWEL strives to match and even exceed such curricular goals. There is also an underlying notion that the self-organising learning orchestrated by children is inherently better, more liberating and more egalitarian than in formal schooling. So this approach could imply that teachers should stay away to encourage children towards free learning.

In fact, HiWEL as an experiment has matured considerably in the last few years. It has moved from primarily proving that children can learn by themselves through computers to the how and what of learning. They have also started to focus more on the building of ties with the school, particularly in regard to using the teachers or others in the local communities as mediators in learning. While this can be a promising move for sustainability through the integration of outside interventions, the question here is how can HiWEL control the kinds of instruction that occurs if the teacher becomes involved? What happens to knowledge discovery and knowledge creation if the instruction by the rural teacher is embedded in rote learning –what would this do to the HiWEL kiosk learning spaces and its specific novel pedagogic activities? What trade-off would HiWEL be willing to make to achieve sustainability – and would this in the end, be self-defeatist? And with computer laboratories springing up across schools in rural India as part of the new digital divide investment policy by the State (Garai and Shadrach, 2006), how will HiWEL maintain its distinctive character? This would surely require HiWEL to move beyond the uses of computers as tools of liberation and learning for children and take into account their more diverse applications. With its hard-earned Silicon Valley status, India now regards the computer as a symbol of nationhood...the extent to which institutions are willing to let go or participate in letting go of such "instruments" of power and persuasion may be worthy of investigation.

Private Distance from Public Education

Let's be fair. If HiWEL involved teachers to genuinely participate in the usage of kiosks, it would have been self-defeatist. There is a high likelihood that teachers would have watched and instructed; they would have reprimanded and monitored closely; they would have tamed the spirit. They would do what they do best, convert children to pupils, to disciples of learning. We would hardly see evidence of spontaneous and systematic self-directed learning. Playing games and music downloads would have been supplanted by *Excel* charts and *PowerPoint* usage. Besides, with limited access to computers in schools, usually for a half hour, three days a week, shared with three other children, there is already a superimposed constraint on learning. Schedules would take over passion, and curriculum would dictate the pace of learning.

In the case of the Almora and Hawalbagh kiosk, institutional indifference manifests in complete abdication of responsibility, as persistent disdain and distrust prevails. Suggesting education as "invasive" barely helps HiWEL's case either. There is a reason why Mitra's "Minimally Invasive Education" phrase hasn't caught on. It is not just the lack of incentives to participate with this *free* learning idea that stops schools from stepping in, even if just to nudge children in this direction or to protect the kiosk. It is the "unknown" that they would be participating in that holds them back. Also, it is the perceived "mis-education" that they could inadvertently be participating in that cements their indifference. Indifference, however, should not be mistaken as disinterestedness. On the contrary, schools are deeply "interested" in all matters of "education" as they continue to sieve through for the public what constitutes as learning. And that, of course, is at times, the problem itself.

People in power, institutional power in this case, do not stay on the periphery and, a lot of effort is needed to stay *out* of their peripheral vision. It is the nature of the beast. Schools do not learn to move themselves, they learn to move others. Lave and Wenger's much discussed "legitimate peripheral participation" can be applied here:

> …to draw attention to the point that learners inevitably participate in communities of practitioners and that the mastery of knowledge and skill requires newcomers to move toward full participation in the sociocultural practices of a community. "Legitimate peripheral participation" provides a way to speak about the relations between newcomers and old-timers, and about activities, identities, artifacts, and communities of knowledge and practice. It concerns the process by which newcomers become part of a community of practice. (Lave and Wenger, 1991, p. 29)

This underlines how entities of a lesser status (e.g. HiWEL) than "old-timers" (schools) are often compelled to interact with and participate with established practitioners when involved in the politics of education. In doing so, there is a likelihood that HiWEL becomes the School. This can be seen in the shift in

HiWEL's pursuits over the years as they earn their legitimacy by striving to prove that they can do what academic institutions do – scholastic achievement, computer literacy and more, all deeply desired by the statehood of schooling. As HiWEL's goals align with the goals of the State on learning, even "fun" can be a genuine threat to schools. Thereby, HiWEL continues to tease, linger, parasitically so at times, on the fringe of schooling.

States are not disinterested in computers as an artifact. In India, the computer has supplanted alternative images of the nation – that of the malnourished child, the sacred cow, the slow moving elephant (trying to outrun the Asian tiger), the yogi on the mountain and more. How can they let go of their grasp of such a symbol in the name of "fun?" Computers have shown the pathway for the new citizen, the new netizens, as they gear up for the next era, not just to furnish the back-offices of multinationals but to be at the forefront of innovation. Thereby, a casual placement of this artifact with no overt agenda outside the public school is naturally then seen as a fundamental problem.

Placing kiosks in playgrounds of schools does not and cannot wall out schools and its schooling structures, expectations and intentions. By stationing the kiosks in playgrounds, there is hope that it will be close to but not part of formal education – free within close quarters. But schools, like temples, are loud representatives of their religion, leaving the sacred compounds to diverse spaces, both public and private. In other words, schooling is a traveling framework…it moves as you move, and you, in turn, help it move.

We need to separate HiWEL as an experiment and HiWEL as an institution. As an experiment, it serves its purpose. It reminds us of the ingenuity of children with their deep and diverse capacity to self-instruct, particularly in the realm of play, a point which is often lost in the maze of schooling. It serves another purpose: it facilitates the revitalization of the public sphere of education. Schools while public, are agencies of the State and within this public persona, lays private interests. Thereby, schools look out for their own interests, sometimes representing not the public but themselves. To keep schools truly public, there is a need for ideas like HiWEL to make public education uncomfortable. It can remind us that education is a public sphere and political, where "consumers" are also "critics" (Habermas, 1989). HiWEL's institution, on the other hand, driven by the need to constantly seek legitimacy (and funding) from the State, while building its identity as an alternative learning space and pedagogy, is in need for reinvention.

Playground Kiosk Democracy

> Accidental or incidental discoveries if repeated within a collective environment leads to learning. (Mitra, 2000, p. 221)

We should be wary of collectives. A thin line divides a community from a mob. There is no averaging of human desire, human potential, and human need that makes a democratic public. Instead, dynamic asymmetries in relation with people, things and spaces is the nature of the game, "…it's all there ever is" (Lave and Wenger, 1991). Human interaction is a series of political acts that is in constant motion. In the process of reproducing the "rules of the game," there is room for reinvention, revitalization, and resistance. The very process of understanding what the "rules of the game" are, interpreting and coming to a consensus is a political and *situational* act. The space within which these acts take place is relational. Just as these informal learning events serve as "ceremonial acts," such communal spaces ends with the ceremony ending (Redfield, 1955, p. 6). However, it's what happens *during* that ceremonial moment which is of interest here.

HiWEL follows a constructivist Vygotskian (1978) approach to learning. They espouse peer-collaborative learning as the root to active construction and invention of ideas. "Learners determine their own learning outcomes," they state (Mitra and Rana, 2002, p. 222), pointing to the spontaneous emerging collectives of children as they are drawn to the kiosks – touching, feeling, fiddling, surfing, nudging, and more, the artifact and its space. And, better yet, placing such kiosks in playgrounds is not just a practical strategy to attract children, it's a symbolic one. It makes the statement that learning with computers in this way *is* free learning, learning *is* play, and play *is* possible by all children, and accessible to all on such public grounds.

HiWEL is not alone in celebrating collaboration and play in learning. Much research has gone into demonstrating that within play, self-structure and self-motivation for learning is embedded (Butler, 2008; Opie and Opie, 1969; Sutton-Smith, 1979, 1997). Rather than labor individually, HiWEL sees, "collective learning efforts" as more natural and rational for children as they play together with the computer and share ideas and strategies for learning. There is a recognition that such strategies emerge through the formation of collectives amongst children where, "children form their own social networks at these learning stations, which facilitate information to percolate from the perceived leader(s) to various learners" (Dangwal, Jha, and Kapur, 2006, p. 296). Such self-organization entails groups of children organizing themselves into leaders, (experts), connectors and novice groups. Here, the leaders and connectors create linkages with other children and disseminates their teaching to all. Underlining this is the assumption that such an endeavor happens cooperatively and learning is shared across the board through such strategies and networks. HiWEL goes as far as to state that this democratic learning circumvents barriers such as age, caste, class and particularly gender, "there is also no gender restrictions as there may be in certain social situations" (Mitra, 2005, p. 80).

However, evidence from HiWEL's own experiments suggest that often there are far less girls than boys accessing these kiosks (Mitra, 2003). My own experiences in rural Andhra Pradesh, South India, for half a year, where Hewlett-Packard set up computer kiosks for the community, witnessed primarily boys flocking to play car games and not much else (Arora, 2005).

Figure 6.3   Boys at a Computer Kiosk in Andhra Pradesh, India

Boys tend to mark these kiosk spaces quickly as "play stations" and these markings become common knowledge over time (see Figure 6.3). Often, the same group of boys dominates these spaces. The point here is to not dwell on sex inequality in computer usage or gender dynamics with technology, where girls are perceived as marginalized and victims of male-oriented technologies and its spaces (Solomon, Allen, and Resta, 2003; Thurlow, Lengel, and Tomic, 2004). While not meaning to discount the importance of this gender discussion, the issue at hand is that "collaborative" learning is not necessarily democratic, and particularly amongst children, not egalitarian. In fact, peer collaboration may not necessarily improve learning and may at times, have a detrimental effect on educational processes and practices (O'Donnell and King, 1999; Tudge and Winterhoff, 2006). Spaces such as playgrounds, located as "free" for collaboration cannot be disassociated with such social practice. In other words, HiWEL's unique association of computers with playgrounds can breed not just collaboration but competition and discrimination in learning and teaching. Here, schools need to be recognized for what they represent – free and equal access to knowledge *by* the public and *for* the public, a genuine and universal accomplishment through centuries of struggle, of making public that which has been for the most part, in the private realm.

Besides, playgrounds are a social construct, the deliberate makings of a public and separate space for children that came to being at the turn of the 20th century

(Chudacoff, 2007). It stemmed from the notion of controlling children by providing a safety valve for their "idleness." These alternative spaces were meant to serve particularly immigrant and working-class boys, who would otherwise be lured to far less "moral" spaces such as pool halls and penny arcades, as in the case of the United States. Within such playground spaces, boys and girls were segregated and supervised closely. The intent was to instill virtue through play, to extend the arm of morality, by gifting freedom with control. In other words, playgrounds were a manifestation of the benign dictatorship of adults over children. To this date, playgrounds, while evoking a sense of free movement for children, are often spaces of acculturation and socialization for children, a laboratory of emerging behavior to be shaped. Interestingly, playgrounds in schools have become one of the few spaces within which children can interact with their peers on their own terms with minimal adult supervision. Here, children freely exercise their choice to discriminate, to decide who to talk to, when and for what purposes (C.H. Hart, 1993; R.A. Hart, 1979; Pellegrini, 1995). Such choices can come with unwanted behaviors of cruelty, distancing, and bullying, often as a direct consequence of competition in schools.

Hence, children can be highly discriminative, where they can ruthlessly eliminate members from their networks through their own "rules of the game." Instead of looking at playgrounds as public spaces where children can be free from schooling structures, we need to view them in relation to their own constraints and markings. These spaces serve as one of the few arenas where children can discriminate freely as they go about selecting who they want to socialize with, when and for what purposes. Hence, these contexts can be seen as potential accomplices for discriminative learning, where play becomes privilege amongst certain networks through a process of elimination and control.

A Beautiful Idea

Free learning is a beautiful idea. In conversation with a high level ex-member of HiWEL, who was instrumental in executing this project at the nascent stage, we can get a sense of the radical nature of such a vision and its honest limitations. This conversation should not be viewed as an institutional defense. Rather, it is a rare opportunity to gain insight through a very candid and genuine practitioner and development veteran:

> Ex-HiWEL: I know many people thought HiWEL was about free education because I had to talk the ones who wanted to try it in their home towns out of their illusions. The way HiWEL actually came about was that the World Bank wanted a cheap, universal solution to end the digital divide, and Sugata, who is a physicist and had experimented with the Delhi kiosk, saw a way to apply Chaos Theory, and particularly self-organizing systems, to education. Chaos Theory basically states that chaos tends towards order by spawning self-organizing

systems, while order tends towards chaos by spawning complexity. It applies to matter, energy, cocktail parties and even you and me. Take India, for example. Since the 1970s, coalition politics has tended towards larger and longer lasting self-organizing coalitions, resulting in the recent decisive vote of confidence in Parliament in favor of the UPA. Therefore we can say that India is tending towards order and stability. But in the US we see increasing complexity and fractures in the political system, so we can say that the US is tending towards chaotic breakdown.

Researcher: So how was the self-organizing system supposed to play out at the ground level?

Ex-HiWEL: You see, self-organizing systems require an "attractor" and at least three "players." In the case of HiWEL, the attractor was the kiosk and the players were the kids. It didn't work if a kid was alone or paired up with one buddy. There had to be at least three, and the more kids the better it worked. The attraction of the kiosk motivated the kids to band together and figure it out. We even had kids who were blind or otherwise disabled doing well on our tests because the group learned as a unit by cooperating. Similarly, MIE worked best in the most remote villages because the kiosk was the only game in town – a powerful attractor. So in reality HiWEL was just as structured as classroom education, only the structure was invisible to the kids. Moreover we monitored the kiosks from Delhi and adjusted the difficulty of the educational games to the learning rate of the kids just as a teacher would. The advantages of HiWEL over classroom education were that it was scalable, encouraged kids to explore and figure things out for themselves, and was language-independent in that the kids all ended up learning English because of the English games and Word, which for some reason was a big draw for all the kids right from the start. Because our parents and kids valued English as the route to upward mobility, we even adapted Dragon's Naturally Speaking to work as an English pronunciation coaching system. So our model did meet the World Bank's criterion of universality and might even have been cheaper with more time and tweaking. We eventually did attempt local languages with our cartoon parrot "Hara Tota," because she had to be able to teach kids in each state how to sign up for email. But our goal was to educate the local content developers of the future, not presume to fill that role ourselves.

Researcher: So what were the results of the scaling of this idea?

Ex-HiWEL: HiWEL was independently replicated by CSIR South Africa in two village locations and Pretoria, where they got almost identical results to ours. Sugata also put up five kiosks in Cambodia, two of which did not work well for similar reasons to Almora. In these instances the kiosks had to be put up in a hurry in a strange location and we failed to provide the basic requirements of self-organizing systems. We had problems with bad sites and older kids who were taking computer classes nearby and chased the little kids away so they could email or visit porn sites. Sugata developed another model for older kids with commercial multi-user domain games as the attractor but not enough gaming machines for all the kids to play. The ones waiting their turns could pass the time by using the computers with educational game content. However, we never got the chance to test this out.

Researcher: So what were the other reasons for it to have worked in some places and not others?

Ex-HiWEL: Most of the kiosks in India worked well because we used local NIIT affiliates familiar with their areas to help us site the kiosks and identify qualified researchers. In some cases like Hawalbagh we had problems with teachers who felt the kiosk had usurped their rightful role, but in many others the teachers were flattered to be sought out for advice by their students and happy to be "guides on the side." As far as locating the kiosks, we used the AT&T principle for targeting telephone booth sites – that they should be in "public view" in areas well frequented by parents, teachers and other responsible adults. We did site some in schoolyards, especially towards the end because we were interested in the reactions of teachers. To be honest, I think that adults are frequently afraid of children who are empowered by technology. We certainly saw this in the Hole, and I think that's probably why it has not been more influential. It was pretty scary for us too at the beginning, especially when eight year old Rajinder broke the news to a BBC film crew that General Musharaf had already left town and the promised peace talks had failed. That was just part of his world, along with the village in Bihar where he was born and the illegal shacks behind the NIIT corporate offices where he lived with his parents. Like the famous headline said, Rajinder was a netizen. That article made him a global celebrity and got the chairman of the World Bank to make a pilgrimage to Delhi and write us a check for two million bucks. Now that's empowerment! I don't know how many times I've explained this to other projects that didn't want to share the glory with the people they were supposedly trying to help, or didn't want to admit that Indians can solve their own problems, but instead accused us of hype and some sort of inside track on funding.

Researcher: So what do you think of what happened with the kiosks in Almora?

Ex-HiWEL: I understand your focus on Almora and agree that the two kiosks there failed. In fact it was the worst failure we had. We used to joke about having two public toilets as our epitaph in Almora, but they shouldn't symbolize the entire project, which on the whole helped a lot of kids in India and elsewhere and added lot of new information about open group learning to the educational record.

HiWEL as an experiment is an important initiative. It has evidenced the ingenuity of children and their capacity for self-learning through play and experimentation, something which is all-too often lost in much traditional schooling in India. It has shown that it has the potential to provide educational opportunities for those denied formal schooling, enhance and extend formal schooling, and remind schools of their purpose and duty to the community. It has even shown that children can be the "pundits" of the new digital age.

However, HiWEL's connotation as an "experiment" denotes its temporary status. And now, while advocating and demonstrating a means of free learning, HiWEL is also becoming more institutionalized, more structured in its design and operations, and more reliant upon the use of mediators to assist the children in their learning. The admission of more formal means of teaching and learning into its informal spaces and processes is needed for the purposes of funding, efficacy and social acceptance. This comes with some significant challenges of negotiating the relationships with the school, the teacher as a mediator, and the kinds of content, instruction, and curriculum that it can allow to seep into these relatively free spaces without compromising on the underlying tenets of innovative pedagogy.

Also, just as in society writ large, learning involves competition as well as collaboration. Autonomous learning with computers without some monitoring and mediators may continue to face resistance both on the grounds of ensuring educational order and that all children have equal access and opportunity for computer access and learning. However, the intervention of mediators within such free spaces of learning can serve as a double-edged sword. Mediators require compelling incentives to intervene, and come with a range of perspectives on instruction and content. Hence, for HiWEL to sustain their vision, they would have to "school" such mediators to their ways of thinking. This could possibly position HiWEL against that of State schooling, making this a game of trade-offs.

PART III

Computing and Cybercafés

Chapter 7
Copycats and Underdogs of the Himalayas[1]

Cybercafés as After-School Centers

Across the silence of HiWEL's Learning Station in Almora town, a din of voices can be heard from recently mushroomed cybercafés. The coming of broadband in this area has triggered euphoria for cybercafés, from four as of last year to around 20 today and growing. Most such spaces are not "new" – they are STD/ISD (telephone) booths, three-star hotels and *chai* shops doubled up as cybercafés in true entrepreneurial fashion. With this new "computer craze," as one café owner remarks, comes a flock of students to patronize these spaces.

Students have supplanted tourists as the prime clientele, as they go about transforming these cybercafés into *after-school centers*, personalizing these spaces to their convenience. For less than 20 rupees [50 cents], students come here to complete their school work, ranging from accessing information/visuals for school projects, typing their thesis (primarily in English), discussing their projects to applying for further education online. For an extra five rupees, they are provided with an assistant to aid them in their tasks. While students serve as the main source of income, they do come at a price. They take over cybercafés, ask questions and demand constant and continued assistance, disrupting all other activity within these cafés that barely have standing room. With limited space at their disposal, café owners squeeze in computers, printers, Xerox and fax machines, music speakers, and a host of technical paraphernalia. Amongst this digital bliss sits the owner and his assistants, ready to serve their demanding clientele (see Figure 7.1).

These cafés position themselves as multipurpose portals, envisioning needs of pensioners and job seekers in filling out electronic forms, small businesses and NGOs for printing their proposals and booklets, to technical maintenance for nearby schools and shops. But most importantly, they aggressively send out signals inviting young customers, promising them, "the way to get access to the whole world." They demonstrate an acute understanding of student needs as manifested through signboards that read, "Software Project Creation With Educational Aspects – You are Always Prominent For Us" (see Figure 7.2).

---

1 Some material from this chapter has been used to produce a case study for educational technology and learning. This case study has been published in S. Marshall and W. Kinuthia (eds) (2010), *Cases n places: Global cases in educational and performance technology*. Charlotte, NC: Information Age Publishing.

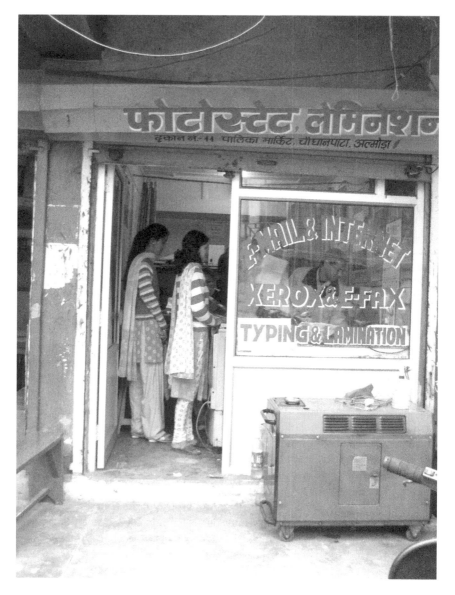

Figure 7.1   Cybercafé in Almora Town

Some café owners, though, choose to opt out of this race: lifeblood to some, a
disturbance to others:

> I prefer to do more maintenance because these students interrupt too much and
> I can't do other jobs like fixing computer parts and faxing and printing. These

children keep asking questions because in school you see, there is less training so they often want to fix this and that and I can't work then. Also there is more software than hardware problems. So I am constantly reinstalling and connecting and fixing viruses. Where will I get time to watch what these children are doing, forget helping them.

Figure 7.2   Signage outside Cybercafé

These "disturbances" are what makes for an interesting reading. What these students do with such spaces and how they do it is of prime concern here.

This chapter exposes interactions of students with each other, the cybercafé staff and others as they go about using computers for accomplishing school projects. This chapter reveals activities that can be perceived as *plagiarism* as students go about actively cutting and pasting texts and overriding copyright, manipulating information and making it theirs, and disguising ownership through collaborative deception. Here, multiple "accomplices," from people to popular websites, are at work. That said, this chapter moves away from the morality of the act, centering instead on the actual processes at work. Also, social learning in its direct and indirect form with computers, digital literacy, the relationship between gender and computers and new technologically-mediated educational contexts are investigated at some length.

You Scratch My Back, I Scratch Yours

In small quarters, each inch of space counts. Students come in pairs or groups, but rarely alone. Café owners have their sidekicks, and new and old technologies play host in such spaces. Upon entering a cybercafé, there is a high likelihood of witnessing the café staff sitting at the keyboard while being instructed by their clients looming over. Students dictate and the staff member types. Students command and the staff member obeys. Even in ignorance, students demand results and the staff delivers. Thereby, to justify and earn my space in this cauldron of "digital" learning, I volunteered to serve as an unpaid staff member for a month at a cybercafé reputed to be the most popular amongst students. Their fame can be attributed to their proximity to the intercolleges and, more importantly, their zeal to cater to student's interests.

This given space hosts three computers, one Xerox machine, two printers, one owner and one assistant. Computers are cordoned off with curtains, typical of cybercafés here, even though rarely are they drawn. Amongst the *Intel* Pentium Desktops, *Philips* speakers, *Sony* CDs and DVDs, *BSNL* Ethernet cards, *hp* LaserJet printers and inkjets, and *Microsoft* software, popular Bollywood songs mark this space loud and clear, coming from a CD loaded on one of the computers. There is seating space for three while the rest are compelled to stand, allowing for constant motion within this 8x10 foot space. On a given day, one can expect the cybercafé owner to be sitting behind his desk, his assistant at the print machine, myself at one of the computer booths, a few students standing and directing me online, perhaps another student accessing their email in the neighboring booth, a friend of the owner hanging around reading the papers, and a *chai* person, usually an old man with his portable tea container, delivering his usual portion of highly sugared liquids every two hours or so.

I offer myself to be instructed by students and wait to be told what to do and how to go about it. I become the learner. It is to everyone's interest that the student succeed in his/her educational task. To enable such collaborative learning, all must do their part. Given that broadband is a relatively recent phenomenon, the owners themselves face a learning curve with the Net, figuring out its potential through IT crash courses, foreign tourists, friends, and students themselves. I am another "source" of learning for the cybercafé staff. Competitiveness between cybercafés is driven by how well the owners are able to extract computer know-how from their customers, particularly those from the outside. Says one cybercafé owner:

> Luckily the tourist season and the exam time don't happen at the same time. The last time some people came here, from I think they said Holland, they would come every day to check their email. This one fellow was very good and we became friends. He showed me that this Wikipedia gives all this information and YouTube and things like that but when he left I typed "Utube," [he writes it down] nothing came. Anyway the broadband cannot handle it [videos] now but Wikipedia is popular with the children. My friend also helps. He has a maintenance shop and sometimes he comes and asks me for help too.

The owner has to be able to share his new computing knowledge with his staff and be able to connect the right kind of learning to the right kind of tasks required by his clientele. Correct technique though does not suffice. Speed is of essence as students operate on deadlines, not just academic but also domestic, as many live in hostels or have to walk a good distance to go back home before sunset.

Take the second year biology student for instance. He has to document the behavior patterns and habitats of animals of a region. He comes in and informs the owner of his assignment. The owner instructs his assistant to get to work. The student sits next to the staff member and quietly watches him as he opens the browser and gets onto the English version of the Wikipedia site. He enters "elephant" in the search section. He highlights the results by selecting all the data, then copies and pastes it onto a *Word* document. He proceeds to another animal. The cybercafé owner comes by and asks the student if he has any restrictions on page printouts and the number of documented animals. The student shares more details of the assignment, stipulating the limits as well as his budget on printing. Based on the given information, the owner (O) instructs the assistant (A) in front of the student (S):

O: Just copy the text, not the visuals and just the first paragraph of each section of these animals from Wikipedia. You should pay more attention when copying. Save some time by not copying all the text, just copy what he [S] asks for.

ओ : सिर्फ टेक्स्ट की कॉपी कर दीजिये, विजुअल्स की नहीं और विकिपीडिया से इन प ुओं हर एक खण्ड के सिर्फ पहले पैरा की; कॉपी करते समय बहुत ध्यान से कीजियेगा। पूरे टेक्स्ट की कॉपी न करके थोड़ा समय बचायें, सिर्फ उसी को कॉपी कीजिये जिसके लिए वह (एस) कह रहा है

A: But sir I was going to get rid of it anyway afterwards.	ए : लेकिन सर, मैं इससे पीछा छुड़ाना चाह रहा था, छोड़िये बाद में
O: What's the point? Unnecessarily wasting time. Do Ctrl C. Why are you going to Edit?	ओ : इसका क्या औचित्य है? बेमतलब वक्त बर्बाद कर रहे हैं; कन्ट्रोल सी करो। आप एडिट में क्यों जा रहे हो?
S: I want different…some new animals for my project, not the same one as the others.	एस : मुझे इससे अलग, अपने प्रोजेक्ट के लिए कुछ नये जानवर चाहियें; बिल्कुल औरों जैसा ही नहींओ
O (to A): Er…and make sure that you erase all signs of Wikipedia from the document, understood?	(ए से) : ठीक से...पक्का कर लेना कि तुमनें डॉक्यूमेंट से विकिपीडिया के सारे नि ान मिटा दिये, समझ में आया?

The staff member asks the student what other animals he wants besides "elephant" and "monkey" in his project sheet but the biology student is unable to come up with any names at the spur of the moment in English. Even the assistant appears not to know. He goes to Google and types "animals," and from that, scrolls all the way down to "animal list" and clicks on it. The list, however, appears to be too "difficult" and "exotic" to understand as they both read through the list and struggle with "parma wallaby" and "coyote" as potential candidates. He goes back to Google search and selects the next choice, from Wikipedia under "list of animal names." He clicks on that link. The student approves the list and from that, the assistant chooses "panda" and "squirrel" for further copying and pasting.

Consequently, the biology student, without touching the keyboard, has successfully accomplished his task by knowing what to navigate (not the Net but the cybercafé space and staff), knowing who to communicate with and when (the owner for more tailored service) and knowing when to stay back (as with the assistant where, rather than actively manage the search, the student receded to the background, getting his money's worth) and when to come forth (expressing a specific request for "new" animals for the assignment). This exercise of fulfilling a task with minimal overt labor is a learning achievement in itself.

It would be simpler to interpret cybercafé spaces and their activities by contrasting them with formal educational institutions, as alternatives spaces for learning away from schools, or, to look at these sites as anti-educational, perpetuating "mis-education" through active plagiarism. However, that would belie the schooled character of these cybercafés and assume institutional ignorance of the goings-on that transpires within these spaces. After all, the accomplices do not end with the cybercafé staff. The situation gets further complicated as we see some teachers demonstrating awareness through their lack of surprise when encountering student's activities in cybercafés.

For instance, two girls (G1 and G2) are at the computer trying to figure out how to access some "good Western artists" from the Net for their mid-term art history exam. At this point, their art history teacher (T) enters the cybercafé. The girls wave him over:

G1: I couldn't find good Western artists sir.	जी 1 : सर, मैं अच्छे प‍ि चमी कलाकर नहीं खोज पाई।
T: What is Van Gogh do you know?	टी : तुम जानते हो कि वान गॉघ क्या है?
G2: He's a painter.	जी 2 : वह एक पेंटर है।
T: Of? Oh, I mean which village is he from?	टी : ओह... मेरा मतलब है कि वह किस गांव से है?
G2: I don't know, from Europe?	जी 2 : मुझे नहीं पता... यूरोप से?
T: So put both his name and his village in Google and search, then you'll find something more.	टी : ऐसा करो उसका नाम और गांव दोनों को गूगल में डालों और ढूंढो, तब तुम्हें कुछ अधिक जानकारी मिलेगी।

Having stated this, the teacher heads to the printer to make copies for the next class assignment. In fact, if one is tempted to see this as an aberration than the rule of the game, focused group interviews conducted with school staff of neighboring intercolleges reinforce the absence of institutional naïvety on this subject. For instance, the interview with a Principal (P) and her staff (Teachers T1 and T2) of an intercollege about this matter reflects this proposition:

P: The computer has invaded our life! What I am concerned about is the impact on students. How they waste away their time on the Net and do nothing useful educationally.

R: Do you monitor them?

T1: They are not doing it here [school]. It's all happening in these cybercafés. They stay there for hours and god knows what they do.

T2: Sincerely they use computers when they have to do projects.

T2: As an information device, they are using it very well but as far as integrating it into their studies, I don't think they are using it well at all. What they are basically doing is getting the information, getting the printout, compiling it and of course, sometimes they have to get some sampling.

R: Do you think they are creating any original work?

T1: No no no. They are compiling it, they are editing it, they are doing some work of their own but that is very very rare. They are producing the information that is already there.

T2: It is not possible the way they are working. They are not producing anything new.

P: They have become very good *copycats* [emphasis added]. That they do very well but if you say do something on your own, they do not know what to do. Original work is missing.

T2: Say if our assignment is the use of various plants in Kumaon and its medicinal uses, they will take something from here, something from there and then they just print and attach it and give it to us.

R: So how do you change this?

T1: If we chose some rare topic then they will give us something of their own because they can't get it from the Net.

T2: Use of computers is not there except for typing purposes. They just put some visuals and text and then give it to us.

P: One day they will be in the global place and they will have to learn the hard way. We will not be there to tell them to get curious and be original, it has to come from them. They need to be motivated to do something creative.

As we have seen in Chapter 6, learning can be collaborative as well as competitive and discriminative amongst actors within particular contexts such as playgrounds. In the case of the cybercafé, a vast amount of cooperation ensues for the student to succeed in what would be viewed by many as "plagiarism." This is much like McDermott's Rosa situation where a teacher colludes with a young girl during a reading session, helping Rosa to not read (Tedlock and Mannheim, 1995). The teacher, with the best of intentions, collaborates to keep the reading momentum going in her class, by not acknowledging the fact that Rosa cannot read.

Therefore, even non-interjection is considered active participation as it takes initiative to remain inactive to the happenings of education around us. In other words, in this act of joint performance, a different kind of learning can be seen, moving away from "learning about" to "learning to;" from content to process based. Here, learning is about how students draw upon their resources by identifying them, communicating with them and bringing them together. With this act in motion, knowing when to stay back, when to come forth, when to interfere, interrupt, and interplay, can be seen as a dance movement to an overarching tune. Surely enough, as the tune changes, so will this meticulous arrangement. After all:

given that no act is an act of isolation, tremendous social effort goes into organizing the politics of teaching and learning, of deciding which moments to pay attention to, moments to test oneself, of one another and the group as a whole; which moments to pay heed to authority, to play to the expectations of the authorized, to succumb, resist, transform and circumvent the rules of the classroom. This complexity of action involves learners posturing according to the task at hand, carefully monitoring each other and themselves, appearing effortless in their (a)synchronized behaviors as they go about relating to one another, and only sometimes judgmentally so. (Arora, 2008b, p. 13)

In organizing the politics of teaching and learning, we need to recognize the inherent instability in relationships between actors and networks as they require repeated performance to maintain its order (Latour, 2005). So, for the student to continue to get her school project accomplished at the cybercafé, actions need to be renewed and reproduced. Deliberate ignorance from the school, assignments that allow for plagiarism and the competency of the cybercafé staff, facilitate this process. Much of this performance can fall apart or transform as part of the everyday interaction. However, there is a certain reassurance in larger stabilizing forces and actors at play (but by no means static or permanent) such as State-mandated curriculum, availability of computers and "relevant" websites like Wikipedia and Google. As we can see, computers become an accomplice, providing a world for the student to inhabit. These objects are not attributed with intentionality but, rather, it is in the association with objects and humans wherein intentionality resides. Through such "mediations," these relationships are repeatedly performed so as to maintain their order.

This ritualization in the mechanics of interaction (the student comes to the cybercafé to do her school project and the staff fulfills the request attentively and unquestioningly), rests on the acknowledgement of possible conflict (where, for example, the staff could start to preach "honesty" in academic work, a teacher could change the assignment to content that is locally-based or set the grading structure along lines of "originality," a cybercafé staff could forget to erase "Wikipedia" from the assignment triggering the necessity for the school to confront this proof of fact; the Net could shut down, so on and so forth). The focus, therefore, is less on how these orders come about and more on how students, staff, teachers, computers, printers, and other "actors" perform their respective roles in the reproduction of education. This audition for actors within this "community of practice" (Lave and Wenger, 1991) deliberately opens up computers, not as pure objects, but what Latour calls "quasi-objects," as natural and cultural collectives, gaining a legitimate voice in this "parliament of things:"

...what does it matter, so long as they are talking about the same thing, about the quasi-object they have all created, the object-discourse-nature society whose new properties astound us all and whose network extends from my refrigerator to the Antarctic by way of chemistry, law, the State, the economy, and satellites

...However, we do not have to create this Parliament out of whole cloth, by calling for yet another revolution. We simply have to ratify what we have always done...Half of our politics is constructed by science and technology. The other half of Nature is constructed in societies. Let us patch the two back together, and the political task can begin again. (Latour, 1993, p. 144)

As in the case of the playground in the previous chapter, where much effort has been made to erroneously view such space as disengaged from the school, we should cautiously tread here, as schooling enters cybercafé spaces in a multiplicity of ways. And through this "schooling," the family affair of social learning transpires.

Who's the Boss?

A girl (G) looms over the male staff (S) at the cybercafé. She instructs him to transfer the text highlighted online from the "Great British Poets" website to *Word* for her assignment, "The Critique of Poetry." She stands while he sits and types at the keyboard. Her friend is talking on the cell phone by her side:

G: These words are not right, correct it. Here, now separate this, these words are joint see. [she remarks as he types onto *Word*.]

जी : ये भाब्द सही नहीं हैं, इसे सही करो। यहां, अब इसे अलग करो, देखो, ये भाब्द मिले हुए हैं।

S: Yes but I will do that later.

एस : हां, लेकिन मैं वह बाद में करूंगा।

G: No no, do it now or you will forget. God! Why are you making this so big? Oh, why are you...nothing...okay continue. [exasperated about the font.]

जी : नहीं नहीं, इसे अभी करो, नहीं तो तुम भूल जाओगे; हे भगवान तुम इसे इतना बड़ा क्यों बना रहे हो? ओह तुम ये क्यों... कुछ नहीं. .. ठीक है आगे बढ़ो

S: I'll go ahead then?

एस : तो मैं आगे बढ़ूं?

S: Ah, okay fine...array, I'll beat you...stop showing me the same photos for this. Okay, the one you showed me earlier will do, don't waste time now. [She finalizes on an English countryside image from Google to compliment her text.]

एस : ओ.के. ठीक हैं.... अरे मैं तुम्हें पीटूंगा.... मुझे इसके लिए वही फोटो दिखाना बंद करो. ...ठीक है, जो तुमने मुझे पहले दिखायी वह चलेगी, अब वक्त बर्बाद मत करो।

Her assignment is almost complete. The assistant is called to attend to someone briefly. He excuses himself. The girl takes his place and sits down before the keyboard. She starts to fiddle with the mouse. She looks at the English countryside image again, clicking back to another image to compare. She looks to see if the assistant is coming back. He is still busy with another customer. Her friend nods in approval while still on the phone. The girl clicks on the image, copies and pastes it onto the *Word* document opened. She also creates a file and names it "English Nature," and pastes the image in there. She goes back to the *Word* document and works on positioning the image with the text. The assistant comes back. She gets up and he sits down. She instructs him to center the image. He uses the space bar to do so. She tells him to align the text to the right. He hesitates. She makes an exasperated noise, takes the mouse from his hand, copies the text, goes to the "Align Right" button on top and clicks on it. She then hands the mouse to the assistant again, continuing to instruct him on typing the title for her paper.

If this was a snapshot or silent movie, we would see a girl standing while a man sits, a girl looking on as the man types and navigates on the Net, a girl getting up from a chair to allow the man to sit at the keyboard, a girl looking back to see if the man is watching as she operates the computer. Images of fear and passivity would prevail. This would feed into the wealth of literature on women and technology: on the lack of equality in access to new technologies as cultural, economic and social barriers come to the fray (Cockburn, 1991; Cockburn and Furst Dilic, 1994; Cockburn and Ormrod, 1993; Wajcman, 1991), the active resistance by women as they self-regulate their access to new technologies; such tools that are indoctrinated as being "unfeminine" (Hafkin and Huyer, 2006), and the gender-divide with technology seen as an extension of centuries of male subjugation. Some blame such divides on the omnipresent media representations of women as home makers prevailing over, say, IT savvy professionals (Harris, 1999). Others blame this on technology as designed and constructed for and by the male, biasing usage (Henwood, 2000). As we shift focus to less wealthy countries like India, these divides become more pronounced as we witness tremendous disparities between men and women in access to not just technology, but other tools of "mobility" such as literacy, health and employment (Hafkin and Taggart, 2001).

In the last two decades, feminist scholarship, particularly with gender and technology, has come a long way in shifting perceptions of women as victims to those as active agents of change (Arora, 2006a; Kabeer, 1994; Kabeer, Stark, and Magnus, 2008). The emphasis is made on women being "different" in their ways of access and usage with technology, moving away from male usage as a benchmark (Turkle, 1984, 1995). Also, some scholarship succeeds in actively disassociating notions of gender, technology and empowerment as we see new technologies being used against women, such as sex-determination tools used in India (Hopkins, 1998; Varma, 2002). Such studies show new technology as a tool of "oppression" instead of "empowerment." While the above literature demonstrates a pluralistic perspective on gender and technology, what is experienced at this cybercafé of a girl instructing a male of her specific needs through the novel medium of the computer seems to be a veritable conundrum. It sits uncomfortably amongst these popular interpretations of relations between gender and technology. So where do we go from here?

In studying how girls access and use computers, we have to be prepared to let go of the gendering of the situation unless overtly indicated. Rather than looking at how being a "girl" affects and is affected through technology usage, we need to take into consideration all aspects of the interaction to see what is revealed. As such, if being a "girl" when looking at learning with computers:

> ...does not manifest itself through this interaction, it is not an affecting factor in our understanding of that act. In schooling, we presuppose that factors like gender, race, class and other "identities" take over the classroom, and in doing so, we allow for the takeover, leaving behind the actual teaching experience. Rather than starting from that point and reifying that which we believe would impact schooling, we should start by looking at the instructional processes within the teaching acts by students and teachers to instruct our understandings. (Arora, 2008b, p. 11)

By no means should this be looked at as a negation of deliberate and active exclusion of women and girls from the technological realm that does take place, as we have seen on the playground in Chapter 6 and will be seeing more of in Chapter 9. It is acknowledged that persistent institutional, social and cultural mechanisms perpetuate and sustain the distancing of women and girls from new technology as well as the fact that creative struggle ensues as women interpret, circumvent and transform their given situations. Yet, in this case, by forcing the lens of gender onto the situation, we would not just be participating in a reductive process of girls as victims of male systemic dominance but simultaneously be evading that which should be genuinely investigated – the unique learning at play. This can push boundaries of what constitutes as gender interaction with technology, technology as a tool of mobility and empowerment for women as well as the male-female power dynamics that tends to seep through when discoursing about such matters. In other words, these deliberate juxtapositions can broaden the parameters on the relationship between gender and technology.

What is most interesting here is how the girl dominates the male staff in executing her task as they play the role of client-staff effectively. Also, fascinating here is how she teaches him and even demonstrates her needs by taking control of the mouse at strategic moments while, at other times, chooses to actively exclude herself from direct computing. This brings to attention how accessing computers should not always be understood as a privilege. At given times, it can be viewed as laborious. School work in this case, comes under the latter category. And given that it's an economic transaction, it's also about getting ones money's worth. Thereby, competency with the computer lies not just in being able to operate it, but also in executing a task through effective delegation, supervision and at times, micro-management.

This process of self-exclusion, the exercise in learning to manage computers and people who manage computers, challenge the dominant discourse on gender exclusion. Rather than perceive this as gender marginalization or indoctrination, of being left behind, being on the other side of the divide, we can start to look at this as an exercise in the freedom of choice. The choice of deciding when to participate and engage directly versus indirectly with the computer, resides more in the hands of the client than the assistant. The economics of transaction dictate.

It is typical of technology that once they lose their novelty status and become "common" objects, their access and usage becomes less a matter of privilege and more a matter of necessity. When new technologies become old, their user group often shifts along lines of class. The typewriter of yesterday, once a mark of privilege and high culture, may now evoke images of people typing in dusty government offices in supposed Third-World countries. Walkmans, desktops, rotary phones and more have been "demoted" and have lost their celebrity status. Second hand PCs get "donated" to developing countries, while being perceived as e-waste by others (Grossman, 2006). Choice can be tied to a range of cultural and social factors, where old typewriters can be a collector's item or junk; where car parts can be trash or valuable resources to create new parts (Miller, 2001) and more. That said, computers are a recent phenomenon in Almora and broadband even more so. With a short history of less than a few years, this choice of operating computers does not stem from their lack of novelty and hardly from low status. It's not the artifact but what is being done with the artifact within a specific space is what counts.

Hence, this self-exclusion can neither be seen as "mindful disconnection," or the "art of selective disconnectivity" (Rheingold and Kluitenberg, 2006) as that implies a saturation of technology and thereby, a need to exercise "an act of will on the part of individual humans as a means of exercising control over the media in their lives" (p. 29). However, in the act of instruction between the girl and the cybercafé staff, there is deep engagement by the girl as she goes about attending to every act of the assistant. In fact, her heightened engagement allows her to exclude herself in the direct interfacing with the computer.

So, let's situate "indirect" learning with computers in the realm of "digital literacy" and "user-interfacing" to take this discussion further. In acquiring "digital literacy," a common and widely shared perception is that beneficiaries of the computer age, the "haves" from the other side of the "digital divide," are those that directly access the computer (Cohen, 2005). These "users" of new technology, is assumed, not just learn how to use computers but also have access to "relevant" content and develop core competencies needed to succeed in this new information age (Gilster, 1997). Going by the popular definition of "digital literacy" as shared below, we should either disregard the student at the cybercafé as a user of new technology or view her as a highly competent digitally literate person who is able to recognize what is "relevant" content, and be able to authorize and instruct precisely to gain the desirable results. For instance, Lanham (1995) states that to be truly digitally literate, one must be:

> Quick on [one's] feet in moving from one kind of medium to another...know what kinds of expression fit what kinds of knowledge and become skilled at presenting [their] information in the medium that [their] audience will find easiest to understand. (p. 160)

Similarly, Gilster (1997) defines this as:

> ...the ability to understand and use information in multiple formats from a wide variety of sources when it is presented via computers and, particularly, through the medium of the Net. (p. 7)

So, there are two prime issues to deal with: one on what constitutes a "user" of technology and the other, on what counts as learning with new technology?

When it comes to "users," Cohen (2005) effectively outlines the complexities at play in terming someone as a "user," particularly through the economic lens. He states that users are often narrowly construed as consumers of new technology or potential customers for current and future products and services. When interacting with the computer, the user is seen to participate, "in shaping products...to shape the worlds we all inhabit" (p. 15). Thereby, the very act of labeling a group as "users of technology" puts focus on a consumer base, their behaviors, needs and interplay with IT products to further improve on product design for efficient consumption. Instead, Cohen urges us to look at users as people and that "people are many things, including users of consumer goods, but that is never all that they are" (p. 17).

More importantly and appropriate to this case, non-users are not necessarily victims of the digital divide, literacy divide or other kinds of barricades that have triggered a frenzy amongst development experts to launch initiatives to eradicate these walls. Instead, they can be active non-users, those who choose to not directly interact with the artifact unless necessary and can control other users to access and use resources to fulfill their goal. This requires us to know something about "people's tactics for not being users" (p. 17). As to what constitutes a user of technology, we have to broaden our parameters by recognizing those on the "periphery," and realize that the one at the keyboard may not be the key user of the computer at any given point in time. Those that choose to actively exclude themselves may, in fact, be highly engaged and "digitally literate."

Of course, the very concept of "digital literacy" has already started to be problematized. The critique of "digital literacy" as confining and reductive of learning processes, and of turning cultural practices of computing into "compulsory consumption of curricularised and certificated learning" (Lankshear and Knobel, 2005, p. 7), stem from the larger critique of the traditional view of literacy as "autonomous" and inherently "progressive" (see Chapter 2 for in depth discussion on literacy). What is suggested is a more situated, cross-cultural and contextual view of such engagements. However, this still leaves us wanting of an analysis of such unique learning with the computer.

Having discounted the multiple lenses through which we can examine such "choice," that of the girl client indirectly accessing computers through direct micro-management and "bossing" of the assistant, it is perhaps time to look beyond communication and education as key resources to situate this discourse in. Instead, a useful parallel that can provide an invaluable insight into such "bossiness" is that of servitude in India. Anyone who has spent some time in India would have experienced the Indian "servant" class, and witnessed even the middle and lower middle class families partaking in such practices. It is considered the norm for homemakers, primarily women, to have domestic help. Unlike typical employer-employee relations based on mainly economic exchange, such relations between homemakers and servants are complex and interwoven. Due to the servant's entry and deep engagement with the private domestic sphere, they are looked upon as "outsiders," and are continuously undermined to sustain distance and reaffirm class status (Dickey, 2000). Elaborate and intricate rules and regulations structure the micro-practices of the servants, through which, distance is maintained. However, the permeability and porous spatial boundaries of private domestic spaces translate into a constant manipulation of closeness and distance between servants and homemakers. While control seems to lie in the hands of the homemaker, she is often deeply dependant on the servant to sustain her status. High amounts of dependency stem not from the inability to conduct chores but from the need to maintain the status of class through possession and control of servants (Sanjek and Colen, 1990). Intimacy and deep dependency is countered by continued striving for distance. The anxiety of authority results in performances and strategies through language, behavior and other means to dictate the servants. Such relations transcend the domestic sphere and, in this case, may have found their way into the realm of the cybercafé.

Thereby, one can argue that, given that the girl at the cybercafé is intimately bound to her educational project and that the assistant is an atypical, "outside" actor in this educational pursuit, may propel the carryover of relations of servitude. In allowing the assistant to "serve" her, she is able to disengage from her educational "chores." Yet, it is required that she pay close attention to the assistant's actions lest he take over the reins of her education, which is part of a privileged and private world. So, she employs strategies of management that is often reserved for servants:

> What dangers does domestic service pose? Servants themselves represent a dangerous mixing of inside and outside. They transgress household boundaries, which are conceived of both physically and symbolically, by bringing the outside in and by taking back to the outside what properly belongs inside. (Dickey, 2000, p. 473)

Boundaries of being on the inside and outside change, depending on the task at hand and larger contextual factors. As we will see in the following chapter, the assistant is capable of steering the ship and taking control, given his "direct" access to the tasks at hand. Sometimes, the assistant is allowed to take control as long as he continues to follow the rules of the game. As in Chapters 5, Chapter 6 and the forthcoming Chapter 8, intermediaries of technology come in varied guises such as traders, NGOs, State officials, teachers, and adding to the cauldron – cybercafé assistants. While status and class form the platform on which the relation between "beneficiary" and the intermediary resides, such platforms come with permeable boundaries that can be strategically shifted, played and manipulated with, based on the event at hand.

The Perfect Thesis

> Acknowledgement:
> In my attempt to collect and interpret the material for the text of this project report, first of all I express my sincere thanks to Mr X, our subject teacher for his guidance at various stages. I could not have succeeded in completing my work without the helpful console of our Head of the Department, "Dr Y." I am also thankful to my parents and friends whose patronage has been a source of my will to complete this project work. (Sobha Pant, B.V.A –III)

Touching yet typical, Sobha's acknowledgement reflects the norms of what any average third-year college student submitting her final thesis would do. Thank the powers to be and exit gracefully. What is not expected as typical is what has happened prior to this acknowledgement being typed by the cybercafé assistant-myself. Here, I am an unpaid cybercafé staff and another accomplice in the process of collaborative "plagiarism."

Sobha enters the café with three books: *Art* by Clive Bell, *Aesthetics (Philosophy of Art)* by Krishna Prakashan and *Sensation and Perception* by Coren Porac Ward. The title of her final project report is, "Comparative Study of Indian Aesthetics and Western Aesthetics." She rests the books by the printer, pages marked by pieces of paper. She requests photocopies of the pages marked. After getting copies of the different sections of the three books, she sits outside the cybercafé, meticulously highlighting and numbering these Xeroxed works. After an hour, she enters the cybercafé and is directed towards me, poised at the computer. She explains carefully her system: only the highlighted sections should be typed, nothing else. The numbering has to do with the sequencing of the paragraphs – this order should be followed exactly as marked. While there are three separate stapled works, I am instructed to search for the numbering across these works as they are all interwoven. She has written the subheadings on a separate piece of paper under which are the given numbers that allude to the specific sections from the three works. I am told to make these subheadings bold. She wants her thesis printed, bound and delivered the next evening. She leaves. I get to work.

While typing, the following becomes clear. This system is simple and precise. There is little room for any mistake in terms of its order. The typing is laborious and terminates a few hours later. Interestingly, the "client" has displayed confidence in the staff to a point where she does not intend to review the work prior to printing. A sampling of this pastiche can be seen below, transitions marked by shifts in text, from a philosophical to aesthetic to psychological:

> [Insertion of Text X] Art as expression, art as communication, art as enjoyment, in fact art in any and every form is a basic and universal human activity found in every age, every time and every place. Although universal in value, art is closely related to the society and culture from which it springs. Art is an integral and organic part of the growth and maturation of a vital civilization. It can never be separated from culture and society. [Insertion of Text Y] Aesthetics, or the theory of beauty in art and literature, has perhaps been one of the early pursuits of the human mind. As literature and the various arts flourish in a society, the attempt to understand the exact nature and causes of their appeal to the reader or the connoisseur should also finds a place in he thinking mind. A thing of beauty is not only a joy forever but is also an enirtation for ever to explore the reasons for that joy. Since poetry and drama are the earliest arts, it was only natural that the science of aesthetics should be inspire everyone thoughts about early poetry and drama.

While the transition between the two texts are arbitrary yet respectably close in its content, the actual quality of the text, as the reader can see, is markedly different. The latter text is not a reflection of my poor ability in typing but sadly is the actual proofread published textbook material verbatim. So a strange dilemma follows. Not only am I helping the client "plagiarize" by typing the work for her, being silent and thereby, active in this process, but, in taking my job seriously, I feel responsible to facilitate a "perfect thesis." There is a temptation to correct spellings and more importantly, to help make sense of the absolute incomprehensibility of lines such as, "connoisseur should also find a place in the thinking mind" as typed above. With the words bleeding red as *MS Word* kindly instructs for change, there is a natural temptation to interfere, take control, to make this text worthy of plagiarism.

Underlining this is the blaring reality of quality issues with the Indian textbook industry. More worrisome is the learning that goes on with these texts as they get enmeshed with other texts for the manufacturing of the thesis. One wonders whether the student lacks the discriminative capacity for quality text, is indifferent to the disparity between texts, or perhaps, is compelled to use the given texts to succeed in school. In either of these scenarios, what is factual is that such texts are in circulation for academic purposes. What is learnt here has less to do with the student and more to do with the educational institution, of which the student is a part and product. The computer here is merely an efficient medium to facilitate this peripheral *bricolage* of text (Lévi-Strauss, 1966).

In learning the "system," the focus is on the institutionalization of text, namely the business of textbooks in education. The cybercafé within this context provides a forum for exploring that which appears remote – the politics of education. That said, let's stay with academic textbook publishing in India for a moment. In investigating this arena, we may cover some of the dominant debates on content in education, particularly in developing countries. In fact, some argue that, "textbooks stand at the heart of the educational enterprise," stimulating educational change (Altbach and Kelly, 1988, p. 3). Like most post-colonial countries, India can be seen as having undergone a similar struggle in textbook production, including issues of language, affordability, quality, localism and more. For instance, India is seen to have one of the largest academic systems in the world with four million students enrolled in close to 7,000 colleges and 150 universities (Altbach, 1993). India is also responsible for producing the bulk of Third-World educational content, spending about 8 percent of its educational budget on research and development (Singhal and Rogers, 1989). Yet, Indian educational content leaves much to be desired.

Indian textbooks are notorious for their poor quality and readability. Unfortunately, they are often the sole reading material for most students, given the dearth of libraries and schools in rural areas, where a substantive portion of the student base resides (Aggarwal, 2002; Banerjee, Cole, Duflo, and Linden, 2003; Fuller, 1989). Higher educational institutions serve as a guaranteed customer base for the publishing industry. This allows for a sidestepping of "quality" as the real customer, the student, is not catered to directly. In fact, most of the textbook sales in India are made to institutions – mostly schools, colleges and libraries. The politics of education sets in as teachers, publishers, government officials, private publishing houses, both local and international, and editors, work together to mass produce textbooks that *they* deem as necessary.

However, instead of viewing this as a conspiracy of massive proportion between educational institutions and the publishing business to reproduce texts for profit-making, we need to see the situation through the lens of the producers of content for academic consumption, the researchers in higher education. With an enormous teacher base in higher education that is poorly paid, they are still pressured to publish and teach, often in English, a language they are often not competent in (Altbach, 1989; Arnove et al., 1992). This often results in the production of poor quality texts and/or blatant plagiarism of published material. In fact, during my stay in Almora, a tenured local college professor was caught plagiarizing entire textbooks. These stories surface time and again, reminding us that the gatekeepers of text, be it the editors and professors, are often victims of their own desperation.

Altbach claims that this kind of institutional behavior stems from the fact that most colleges and universities in India were designed to be mediocre during the time of the British colonizers, with the explicit intent to produce human capital that could perpetuate order and not creativity or originality. After independence, he laments, there has been little genuine effort to systematically change such structures. To substantiate this, we just need to look at the fact that, with a diverse linguistic population that speaks about 15 principle languages and about a thousand dialects, 50 percent of the current textbooks continue to be published in English (Canagarajah, 1999). One should not be surprised that with computers, the tradition carries on, with most of its online content being in English (Phillipson, 1992).

To be fair, India is not alone in its post-colonial dilemma as English can be a reminder of its colonial past while serving as powerful national glue and a strategic global means of communication. In fact, the burgeoning of private schools in India that are currently a deep threat to public education is, in fact, directly associated with English being the prime medium of instruction (Sen et. al., 1995). In other words, there is a clear demand for English across India, much like computers, as tools to wealth and prosperity. Thereby, Sobha's attempt to collect material for her project is an explicit learning effort to decipher, navigate, and amalgamate texts, both online and offline. In doing so, she can be viewed as a *copycat* who has blatantly lifted texts and made it her own. She is also an *underdog* who is taught to navigate a system which is inherently flawed. Yet this student will most likely succeed in a system that is designed to produce such a "perfect thesis," as gatekeepers of content stand side by side in this endeavor of education.

The "Epidemic" of Plagiarism

"Plagiarism" is the big elephant in the room. In this chapter, there has been a deliberate obfuscation in addressing it directly, admittedly at the expense of the reader's patience. The goal is to not evade or ignore but to compel us to exist with discomfort for awhile. By placing the reader with this "elephant," the intent is to enable us to move from reaction to response.

Exam time in Almora brings out a host of learning events. Amongst these events of collaborative learning, evasion of ownership, and (in) direct interfacing, the issue of plagiarism seems to be the common thread. Accomplices include teachers, publishers, public schools, students, cybercafé staff and I. It's all in the family. While plagiarism is deemed as a high priority issue in the West, it is looked upon here as either an irritant or a problem of secondary significance. Even amongst those who are concerned, this does not evoke urgency for change. This can be attributed to some of the contributing factors below:

- State exam syllabus requires rote memorization and reproduction of enormous amounts of material.
- Dearth of quality and updated library resources available for students, particularly in English.
- Pressure to publish on the overburdened and often under-qualified faculty at the universities.

- Higher Education instruction is in English while the local and first language of students is either *Pahadi* or Hindi, compelling them to seek for "assistance" through the resources at their disposal, including texts online and offline as well as other actors and institutions to help them succeed.
- In this pursuit to complete their projects in the "correct" way, grammatically, stylistically and the like, they resort to "borrowing" texts verbatim.
- Need for students to work together given the limited computer resources available further facilitates this process. (Satyanarayana and Babu, 2008; Tiffin and Rajasingham, 1995)

Hayes, Whitley and Introna (2006) sum up succinctly the prime issues regarding plagiarism within the Indian Higher Education System:

> The standardized syllabus taught to large numbers of students over large numbers of years also means that it becomes worthwhile for entrepreneurial students and teachers to write and publish "guide books" which cover all the topics for a particular course. These books are frequently available from university copy shops or markets and are, essentially, the key ideas from the course and its texts, presented in a simplified format and frequently (though not always) without attribution of where the original concepts came from. In addition, as the intention is to provide students from any college with the knowledge that they require in order to be able to answer the examination, they often focus on a conservative, uncritical approach to the subject. (p. 1731)

So while plagiarism has been written and discussed extensively in the West, it often falls to the sidelines in India. In the West, plagiarism is primarily looked through a moral lens; "academic dishonesty," a "moral hazard" in education, and an "epidemic" of sweeping proportion that needs to be "combated" (Eisner and Vicinus, 2008; LaFollette, 1992; Lathrop and Foss, 2000). Some go beyond condemnation to cultural understanding, attributing such practices to ancient traditions of learning in diverse cultures such as in India or China. Here, reproduction in learning is seen as a strategic cultural means to deepen knowledge and understanding. Further, Asian cultures are perceived as being unaware of their actions when it comes to what is seen as plagiaristic (J. Anderson, 1998; Buranen and Roy, 1999). There is also an economic angle where the steep cost of attending higher education creates tremendous pressure for students, particularly from developing countries, to perform and excel, even at the cost of their academic integrity (Pennycook, 1996).

In this case, contrary to the above, there is a general awareness of the "cut and paste" culture being non-acceptable in schools amongst the students and the cybercafé staff, as they systematically eliminate signs of the source, such as that of Wikipedia, from the assignments. This awareness extends to institutional settings as teachers lament on the current "behavior" of their students, expressing their knowledge of not just the acts but the location for such acts – the cybercafé as a hotbed for plagiarism. In trying to situate, condemn and/or rationalize what constitutes as plagiarism, the enormous accomplishment of computing, by a relatively nascent group of IT users in a small town, in a supposed developing world, can get overlooked. The sophistication in strategizing comes to play as students learn to draw upon the multiple resources available to them – technical, human and socio-cultural and thereby, exercise their authority over the authored. While this is not to romanticize acts of copying, this chapter provides a forum to unearth and question some popular contemporary assumptions in user-interfacing and education literature: collaborative learning as cooperative, indirect usage as disempowering and a signal for "digital illiteracy," and the separation between formal and informal contexts for learning with new technology. While plagiarism underlines these revelations, it is deliberately de-centered.

Chapter 8
Let's Go Shopping!

New Educational Consumers

Consumerism can be seen as the new "participation" in education. The convergence of information and communication technologies (ICT) has marked the past decade as the "information" age (Kenway, 2006), the "knowledge economy" (Drucker, 1995), to the "new world order" and digital revolution in the making (Harvey, 2005). With such promises, the responsibility weighs heavy to prime the youth with a more globally encompassing educational diet of skills and information to help them cope and even excel in the 21st century. The Net is seen as a leveling ground for access to global knowledge. Hence, a reconfiguration of local curriculum has been proposed, spanning across people, places and perspectives (Gardner, 1999). More importantly, there is a focus to harness tools of the new generation – computers and the Net, to enable this goal. Through this approach, "local" knowledge is seen to have a chance to share the limelight with more dominant "global scripts" (Meyer, Boli, Thomas, and Ramirez, 1997) and even overtake, and/or transform that which is considered shared global knowledge (Ginsburg, 2002).

Without doubt, with new technologies come new expectations. As youth lead the way in user-generated content, there is a push to view what users already do online as integral to their learning (Gee, 2007). Specifically, there is recognition that learning with the Net requires a significant shift from text-oriented learning to a more integrated model that incorporates graphics, video and sound (Kress and Van Leeuwen, 2006). In other words, as youth learn with the Net through online browsing, they encounter information encoded in a multiplicity of ways that allow for meaning-making, arguably creating new kinds of knowledge. With each mode, be it an image or a printed word, a unique affordance comes with it. These boundaries are played with by learners as they engage with online material at their disposal.

This chapter takes stock of this thinking by focusing on how learners play with online material and through this process, learns. By looking at a learning event between two girls and an assistant at the cybercafé as they browse through Google to find "Indian" and "Western" painting images for their art history project, their understandings come to the fore as they discuss which images to reject and accept for both categories. This provides an ideal opportunity to ask some pertinent questions: what kinds of learning goes on as these girls interact with such images online? What is the relationship between consumption and production of knowledge and ownership of that online knowledge? What is the difference between knowledge and information? And what constitutes as "global knowledge?"

Shop Till You Drop

Let's look at the following event. The assistant (A) at the cybercafé seems frustrated with the two girls (G1 and G2) standing beside him. It's been almost 20 minutes and there has been little progress in their search. Both girls have been discussing their project while the assistant has been waiting at the computer. Their project is titled, "Western versus Indian painting." The assistant is looking for guidance on what to type in the Google Images search section and deposit into the girl's respective folders. He has already set up two *Word* documents with their names on it as well as two folders on the desktop named after the two girls respectively.

A: So what do you want now?

ए : तो अब तुम क्या चाहते हो?

G1: Type "painting."

जी 1 : ''पेंटिंग'' टाइप करो

A: Too general, what type of painting? Indian? Western? [He types "painting" anyway in the *Google Images* search box. It reveals a range of painting from "custom paintings from photo" of a pencil sketch of a girl sitting down with her head bowed to a picture of a tomato; from realism to animation, all kinds of images come forth.]

ए : बस इतना ही, किस तरह की पेंटिंग? भारतीय.? पि चमी?

A: At least give me an artist name…there are too many paintings as you can see.

ए : कम से कम मुझे किसी कलाकार का नाम बताओ… तुम देख सकती हो कि वहां इतनी सारी पेंटिंग्स हैं।

G1: Array…what artist? We want something new…keep scrolling no.

जी 1 : ठीक से…कौन कलाकार? हमें कुछ नया चाहिये सूची देखते रहो

G1: Not this. Remove this and put it in my file. [She points to the "Portrait of Megan" to be selected, a realistic image of a woman's head.]

जी 1 : यह नहीं… इसे हटाओ और मेरी फ़ाइल में रखो।

G2: Show some more paintings. [The assistant types "Western painting" this time. A host of cowboy images and horse paintings appear.]

जी 2 : कुछ और पेंटिंग दिखाओ

G1: Take this out and save this in my file. [She points to a painting of horses.]

जी 1 : इसे लेकर मेरी फ़ाइल में सेव करो

G1: Put it in my document too... make this fatter okay. [He adjusts the width accordingly.]

जी 1 : इसे मेरे डॉक्यूमेंट में भी रखो... इस मोटा करो। ओ.के.

G1: Not this. [In reference to a picture of a man with a cowboy hat standing next to his horse.]

जी 1 : इसे नहीं

G2: Why not this one?

जी 2 : इसे क्यों नहीं?

G1: No yaar the color is not good.

जी 1 : नहीं यार रंग अच्छा नहीं है

A: I can make the color better.

ए : मैं रंग को इससे अच्छा कर सकता हूं

G1: hmmm...forget it...just this one with those horses.

जी 1 : हूं... इसे छोड़ो... बस केवल यह घोड़ों वाला

A: Give me an artist name. [He mentions again.]

ए : मुझे एक कलाकार का नाम बताओ

G1: I don't know just see what there is...see properly okay. [He scrolls up and down the first page.]

जी 1 : मैं नहीं जानता, जो है सिर्फ उसे देखो... अच्छे से देखो ओ.के.?

G2: Okay I'll take this one. [Pointing to the Native American Indian painting by artist Carl Sweezy. The assistant puts this image in both her file and document.]

जी 2 : ओ.के. मैं इस वाले को लेता हूं

G2: What else?

जी 2 : और क्या?

G1: Put this in the center... what are you waiting for? [Referring to her horses painting.]

जी 1 : इसे बीच में करो... किसका इंतजार कर रहे हो?

G1: Put "Western painting" as my title. Make it bigger yaar, who

जी 1 : मेरे टाइटल में "पश्चिमी पेंटिंग" रखो.. .बड़ा करो यार, इसे कौन पढ़ सकता है?

can read this? [The search carries on. It takes about 45 minutes to go through these images. The girls still do not like any images and have not selected any further images for their files. At the 361–380 image search, "Mona Lisa" appears.]

G1: This one definitely.

जी 1 : निि चत रूप से यह

G2: For me also save this and put this in the front. [First page of her document.]

जी 2 : इसे मेरे लिए भी सेव करके इसे सामने रखो।

G1: How many do we have?

जी 1 : हमारे पास कितने हैं?

A: Two.

ए : दो

G2: Only two?

जी 2 : केवल दो?

G1: Okay go to Indian now. [The assistant types "Indian painting" in the Image search. A mix of contemporary, religious, ancient and classical paintings surfaces.]

जी 1 : ओ.के. अब इंडियन में जाओ

A: Do you want this? [He points to a painting of God Rama standing as Hanuman, the monkey God kneels before him. The girls ignore his suggestion.]

ए : क्या तुम्हें यह चाहिये?

G1: Save this. [She points to a Ravi Varma painting, a renowned Indian painter of the late 1800s.]

जी 1 : इसे सेव कर लो

G2: I want one Varma too but not the same type.

जी 2 : मुझे एक वर्मा की भी चाहिये लेकिन एक ही तरह की नहीं

A: This one with the fruit is good.

ए : यह फल वाली अच्छी है

G2: Er...okay. [He selects and puts it in her file.]

जी 2 : ओ.के.

G2: No, no…take this out and put this one instead. [She selects a woman with the baby painting.]

जी 2 : नहीं नहीं… इसे निकाल कर इसकी जगह इसको रख दो

G1: Take the one with the fruits… the color is better.

जी 1 : फलों वाली को ले लो… रंग ज्यादा अच्छा है

[G2 appears to be thinking about this suggestion.]

जी 2 : लगता है इस सुझाव के बारे में सोचना पड़ेगा

G1: No? You want or not?

जी 1 : नहीं तुम्हें चाहिये कि नहीं?

G2: This one doesn't have a nice face. [Explains on why she is rejecting the fruit painting over the woman with the baby.]

जी 2 : इस वाली का चेहरा सुंदर नहीं है

G2: Make this fatter. [Of the Ravi Varma painting of the woman with the baby.]

जी 2 इसे ज्यादा मोटा बना दो

G1: Too fat.

जी 1 : इतना ज्यादा मोटा?

G1: It's becoming bad now…get it back to the same. [About her friends selection as the assistant plays with the width of the image.]

जी 1 : अब यह खराब हो रही है…. इसे बिल्कुल पहले जैसी करो

G2: Should I take this out? Is it looking nice or not? [Asking about the Native Indian painting to her friend.]

जी 2 : क्या मैं इसे ले जाऊं? यह सुंदर दिख रही है या नहीं?

G1: It's okay..nothing great.

जी 1 : ठीक है…. कुछ खास नहीं

G2: Take it out or not?

जी 2 : इसे ले जाऊं कि नहीं?

G1: Keep it…I guess you need one anyway.

जी 1 : इसे रख लो… मेरा अनुमान है तुम्हें एक तो चाहिये ही

Fatter, nicer, more colorful – these are some of the criteria for selecting paintings to represent sweeping categories of "Indian" versus "Western" painting. This search resembles a shopping spree. The famous Ravi Varma gets accepted on grounds of a pretty face, horses share glory with the Native American as "Western" art, and the Mona Lisa gets the nod from both. It needs to be noted that, while all the images led to hyperlinks on the artists, not once in the entire session did the girls explore these paths. Also, there was no attempt to make note of the artists or the title of the paintings.

This peripheral consumption comes with tremendous effort. Almost two hours and approximately 400 images later, they each settle for two paintings each. They are picky customers. They want their product and they want it right. What is most fascinating is that while Mona Lisa gets selected right away, they exhibit a fuss over other images, not based on whether these paintings duly represent the given categories but on purely personal aesthetic grounds. Hence their "shopping bags" are veritably democratic – big brands like Ravi Varma and Da Vinci are cushioned closely with the obscure Portrait of Megan and Carl Sweezy of Arapaho.

As images get consumed at a rapid rate without much pause, a mix of window-shopping and impulse buying, can we state that this consumption is learning? And if so, what is being learnt? Current literature on consumption has significantly shifted from the woe of passivity to that which is agentive, resistive, creative; reflective of the ingenuity of human interaction, and, learning (Brook and Boal, 1995; Fiske, 1991; Horst and Miller, 2006; Mankekar, 1993). There is less focus on the hegemonic nature of institutional structures and organizations that steer the production of what gets consumed and more on how people interact with such products to create their own understandings. Hence, consumers are producers too as they go about transforming the products of consumption through social interaction.

That said, as these girls surf online, how much of actual "meaning making" is really going on in a substantive way? Barthes (1975) differentiates between the "readerly" versus "writerly" approaches in interaction with modes or "texts." His concept propels us to look at the passive and the active ways of interacting. Herein, the "readerly text" is viewed as a prepackaged entity, allowing the reader to passively absorb the "ready-made" meaning ingrained within it. These sorts of text are "controlled by the principle of non-contradiction," that is, they do not disturb the "common sense," or "doxa," of the surrounding culture (p. 156). Texts that are constituted as "classics" tend to be read in a prescribed way. In that sense, Mona Lisa *is* Mona Lisa, and that represents the "West." There is no other way of seeing her.

"Writerly texts," on the other hand, inspires the reader to shift from being a consumer to a producer of the "text," celebrating the active and creative efforts with a culture and its texts. So Ravi Varma, the painter-prince of Travancore of the early 1900s can be viewed as a million-dollar Sotheby's investment to a highly colonized replica and reminder of the British past. To these girls, it is less about Ravi Varma and more about the aesthetics of the painting. Here, aesthetics is not viewed in the artistic sense that encompasses composition, palette, perspective, and historical significance but as personal taste, the whims and fancies of a momentary preference exercised through choice. The real dilemma comes in deciding what constitutes as "readerly" and "writerly" text. And with all good debates, there is no resolve.

While the girls went through about 400 images for their highly selective two paintings, their time and engagement with each image was fleeting. Gitlin (2001) reminds us that, in many cases, "people do not necessarily 'make meaning' from the images and sounds that they take in or sift through or that sift through them" (p. 7). He points out that there is a disproportionate emphasis on how people interpret, interact and intervene with the world around them while they, "may well be escaping from meaning" itself (p. 10). Perhaps so, as this event appears to reside more on the plane of sensation than of meaning. Yet, we cannot discount the girls' deliberate engagements in search for the "right" images for their art portfolio. Basically, we need to shift from interpreting the engagement *with* these images to the actual navigation *through* these images. Content sits on the sidelines while experience moves through it. Benjamin's (1999) *flâneur* lends an understanding to such modernity of experiences and lifestyles. Much like de Certeau's concept of "walking" as explored in Chapter 6, the *flâneur* is a person who walks the city in order to experience it. More a detached observer, he serves as the tourist, peripherally engaging with his environment. The *flâneur* becomes an expert in remaining a novice to what is around him. There is little congruity or unity around, as contradiction becomes the condition of the surrounding space. This circulation of images creates a space not of permanence and rootedness but of consumption – transience; fleeting moments of experience, and learning.

So if we are to understand these movements through the practices of *flâneuring*, we have only to turn towards Burbules's (2000) work on learning through online browsing to further go down this avenue. He addresses this learning as a journey into space – getting lost, wandering, pausing to look; exploring, asking, observing, listening, rushing through, taking in, lingering, and sharing. The act of movement is what learning is all about:

> All travelers on the Web have experienced disorientation of finding themselves in a strange location, in between the familiar and the unknown, and wondering "Where am I?" "Why am I here?" "What does this mean?" …this sense of being "lost" is an aporia: a problem of having arrived in an unfamiliar location, a riddle of uncertain signification." (pp. 171–2)

While these girls glance quizzically as they scroll through cowboys and Indians and cherry blossoms all on one page, there is a journey to make sense of "Western" painting. They learn to sieve through and discard the "anomalies" of book covers, home photographs and paint brushes that emerge in their search. They learn to exercise their discrimination between erotic tantric postures, religious motifs, festival imagery, Maharajas surrounded by women and *Madhubani* paintings in their "Indian" search on grounds of their own making. Arbitrary or not, these selections form collective representations as they start to weigh in their shopping bags. While the natural layout online is more "rhizomatic" than hierarchical with few preordained rules to determine direction (Deleuze and Guattari, 1976), these girls follow their search systematically, from one set of images to another with the "next" arrow. The journey takes to signposts to chalk the path ahead.

A word though on anomalies in learning: on what grounds did the girls reject the "cherry blossoms" that appeared on the first page of the "Western painting" search and, instead, chose Leonardo da Vinci's "Mona Lisa" after 361 images? Why did both agree readily without discussion on "Mona Lisa" while the others were debated about? How did they decide on Ravi Varma to represent Indian art while deciding against the pre-colonial art of India that would have been equally if not arguably more legitimate? While most readers familiar with these two categories of paintings would immediately agree on these choices, they may be hesitant, if not surprised, at the random selection of the unknown artists and works that were selected to accompany these esteemed artists.

Here, learning can be seen as a spectrum rather than a practice that initiates and terminates as the event ceases. After all, nobody starts to learn on a clean slate. The girls come with knowledge formed through prior complex practices with a range of institutions, be it their home, school, media and more. These "funds of knowledge" (Moll, 1992) come into play as they interact with these images to (re)produce new meaning. Hence, while these girls shop around for "appropriate" images for their portfolio, their prior knowledge helps them navigate through this visual maze. While the "cherry blossoms" clearly stand outside the boundaries of "Western" art, the Native American makes it into the fray. Here, boundaries of this concept are as much constructed as the concept within it.

However, we need to resist the allure of elevating such knowledge to the status of inherently being more "true" or better than formal institutional knowledge. This is tempting to reify with populations that are marginalized from the mainstream by legitimizing what they bring to the table. Noble the intention, the romanticizing of "local" or "indigenous" information (as expounded in Chapter 5), as essentially being more authentic and somehow "better," must be discarded. Education takes place just as much as mis-education as we can see in this event but, either way, learning still goes on.

Mona Lisa and Bathroom Tiles

Mona Lisa did not come easy. Three hundred and sixty-one images later, following the act of "nail art printing," "payment terms at Western Union," and Hong Kong's police force painting and calligraphy club's exhibit of a girl and her dog, Leonardo da Vinci's perhaps most enigmatic and labored work made it as part of "Western" painting. And to rub salt in the wounds, when it was found, it was seen to rub shoulders with the image of custom painting on bathroom tiles. Clearly the organization and order of images through this online search is not aligned with what constitutes as "Western" painting. In other words, few would argue that these two disparate images belong together. In fact a random search for "Western art curriculum" reveals the nuances that barely manifest in this search. Click on any

Western art curricula online, for instance. The web page will be peppered with sections of classical, medieval, renaissance, and baroque styles, early modern to late modern art, aesthetic theory, drawing, sculpture, and architecture. Say we take this further and enter "Western art history key artists" in the search. If it's modern art, there is no escaping the names of Picasso, Klee, Chagall, Dali, Duchamp and more; Rembrandt for Dutch baroque, Leonardo da Vinci for Italian renaissance, so on and so forth. This is not to say that this is necessarily the "right" content to be learnt. Art history, just as any other historical study, is the streamlining of "reality" into a consensual and linear form made possible by those in authority (Zinn, 2003). This kind of structured curriculum is deliberately manufactured as the most commonly shared perception on Western art.

If we are to look closely at some of the images that emerge from the "Western painting" search such as the horse painting, the reader will discover that the artist of this "Western" painting is actually Chinese. In following the hypertext pathway laid out for those who seek further, you will find that the artist, Hsu Pei-hung (1895–1953) is a native of Kiangsu province. He studied painting under his father and then went abroad to Paris and Berlin. After his return, he successively filled various important posts of the painting circles. It is stated that, "he is a painter representing modern China." He is said to have assimilated the traditional Chinese ink painting style with Western thematic influence, and is especially celebrated as a painter of horses.

So here is the wonderful irony of globalization of content through the Net: the consumer is Indian, the producer Chinese, and the product Western. Or can we say that the product and producer is both Indian as the indigenizing of global information makes it local? Let's push this further. The Indian consumer is seeking for "Western" art images for an "Indian" school project; the producer is using "Western" themes with a "Chinese" style to appear modern. Formal institutions in the West are unlikely to allow a Chinese-produced, Western-themed painting to pass as Western art. Yet whether formal Western education systems validate this effort or not, as long as this Western categorization and interpretation is accepted by the formal Indian education system, this perception becomes a local reality. In short, information does not have to be "right" to become knowledge.

In fact, the Geertzean notion that for local knowledge to become a legitimate global candidate, it has to achieve a level of impersonalization appropriate for a universal script needs to be, again, reexamined in this context. After all, this concept continues to be popular amongst several scholars:

> A good part of globalization consists of an enormous variety of micro-processes that begin to denationalize what had been constructed as national – whether policies, capital, political subjectivities, urban spaces, temporal frames, or any other of a variety of dynamics and domains. Sometimes these processes of denationalization allow, enable, or push the construction of new types of global scalings of dynamics and institutions; other times they continue to inhabit the realm of what is still largely national. (Sassen, 2006a, p. 1)

This preoccupation of "global" versus "local" knowledge spawns multiple ideas, one being that "diversity" of information is merely the organization of sameness through packaging and price, strategically shelved to determine human needs/ wants. This gives the illusion of choice. While this is fraught with warning on the sterilization of knowledge through globalization, others view this standardization of knowledge as an achievement in leveling the field for all consumers across the board (Lechner and Boli, 2005).

Here, global knowledge as "world culture" is seen as a transnational and transcultural convergence of patterns and practices where negotiation takes place, "between the particular and the general, between the temporal and the timeless" (Scott, 1995, p. 5). These authors propose that such "global scripts" are often the product of joint efforts of several actors in the process. They are quick to point out that this does not mean that there is a single global script or model defining action. Instead, there is a multiplicity of global norms that guides local practice. Global "curriculum" here, is seen to absorb diversity. In fact, they stretch this viewpoint to encompass partnerships that strive for and espouse "world consciousness" to help generate that which is "global." In this light, world culture is not the sum of all cultures, but rather, a holistic and all encompassing phenomenon.

While these authors talk of a joint effort in the shaping of these global scripts, Meyer et al. (1997) question this harmonic transaction. They argue that in the spread of formal education worldwide, global "consensus" happens through institutional pressure. Being in concert with collective goals of increasing human capital, educational investment, and the spread of knowledge, the likelihood to achieve commonality in curriculum is seen as likely. Hence, they state that since all societies experience the same institutional pressure, global curriculum is likely to become more similar over time.

But the fact remains that, while there is disagreement on whether these "global scripts" are in essence good, necessary or inherently superior to "local scripts," the convergence of information is in need of deeper investigation. The idea that you can actually "negotiate" between "local" and "global" knowledge, that there is a forward and linear direction to which the "local" is moving, that the "global" guides the "local," that these entities are fixed and unchanging or that knowledge can be standardized, can be seen as problematic. In fact, the beauty of the Net is in its allowance to prove this empirically through a mere effort of an online search.

If we are to follow the hyperlinks of several of the images at our disposal through the Google search for "Western painting," we will discover complex and non-cohesive narratives. From bathroom painting to Mona Lisa, there is no hiding of such incongruence. Besides, there has been little attempt to "impersonalize" local content so as to make it "global." The Chinese artist's rendering of horses can be argued to be highly personalized and endemic to this artist's vision of modern Chinese art with Chinese characters marking his painting. Being consumed by an Indian, as in this case, can arguably make this product "global." Hence, the key revelation here is that globalization of knowledge is not in the transformation of content but in its sharing. As Castell (1996) suggests, as society intersects with technology, networks or "flows" become more visible, revealing the mapping of information in their dynamic and variegated forms. So, the focus here is not on how the "local" transforms but on how the range of movement comes to play as it moves through new mediums such as the Net.

This process though is hardly democratic. Between users who have access to computers and the Net[1] (less than 25 percent of the world population with the majority in North America and Europe) and the private sector that often pays to promote their information in this highly competitive online market of content, the "global script" comes through as incoherent or for want of a better word, "eclectic." The online search is slave only to that which is (made) most popular. Here, obscurity wins over popularity if done right; a commercial art gallery in Phoenix gets to appear before a museum art piece in the Louvre if advertised right. Folk art from an Indian tribal area can come before a Husain painting if the Indian government has invested in its promotion. The winner is therefore the best online marketer for her product. Products become information and information becomes a product.

Besides content, the act of browsing through this "global" maze of information brings to light important questions: how far back should we go to categorize something as "Western" painting? Does "Western" painting have to come from the "West?" Does "Western" painting have to come from citizens of the "West?" What is the role of popular versus fine art in such categorizations? How does status of particular people and their art affect what makes it classical "Western" painting? What is the relationship between the painting medium and its art status? Does this search mean to look for that which is "typical" or that which is "classical" to Western art? When we search for Western painting, are we searching for the history of Western art or Western painting techniques and styles? Who decides on such criteria?

By drawing of arbitrary boundaries around a subject matter, it contains as much mis-information as information itself. To convert information into knowledge and apply it, the learner needs to engage in higher-order thinking. While information is often free, knowledge is not. It costs time, effort and inclination to convert random information into that which is needed, desired and is most appropriate for one's goal. Hence, while there is no guarantee that learning will happen, browsing through online content does promise learning if we pay attention to what guides us in our search:

> ...every navigator, every explorer, has had the experience of looking for one thing but finding something else instead; sometimes this new thing is even more useful, more interesting, more enjoyable, more important, than the thing one was looking for...the associations we encounter within a web Do not always make sense, are not natural or inevitable, do not explain themselves. Sometimes we learn by being returned to the same point again and again; with each return, each repetition comes a new recognition, a changing understanding. We also

---

1   Internet World Statistics: http://www.internetworldstats.com/stats.htm.

can learn from occasions to reflect on why webs are designed the way that they are and how they might be designed differently; this both problematizes the apparent naturalness, the transparency of given association and beings to make explicit the processes of design so that learners can create new links, drawing lines themselves and not only tracing them. Here we become familiar with complexity, and complexity's sibling, uncertainity. Curiosity and interest, which are essential to learning, grow out of and depend upon feelings of doubt and puzzlement; they do not threaten interest, but can enhance it. Something that does not puzzle us isn't interesting. (Burbules, 2000, p. 183)

Are Finders Keepers?

While Chapter 7 has focused on addressing learning through "plagiarism," this section touches upon an event that blurs lines between issues of copyright as students mine away the resources on the Net for their projects with little consideration of ownership. This brings forth the recurrent theme in this book of what constitutes the "public" (see Chapter 6), given that youth often choose to reside more within online spaces that promise a social barrier-free experience. However, one needs to be reminded that with "public property," finders are not necessarily keepers. In essence, the characteristic of being "public" lies in its resistance to private possession. In the notion of being public lies, the dual nature of being policed as well as freed. This is the bane of several institutions – educational, media industries and the government, as they struggle to regulate that which is considered by many as the very essence of the Net – its unregulated nature.

Thereby, consumption and production of online resources fall into a similar bind. To better understand the learning that goes on with such "public property," in this case the actual online "content" that students navigate through, we need to enquire into the relationship between ownership and knowledge. So far, we have been dealing with the *how* of learning. In this chapter, we extend the discussion to the *what* of learning. In doing so, we discover how learning plays out as knowledge gets constructed within boundaries that are "globalizing" in ways that are not necessarily from "global" to "local." The above instances serve as an awakening to the range of information that becomes knowledge based on how people draw from their prior engagements with institutions, both formal and informal.

Chapter 9
Leisure, Labor, Learning

Orkut Saves the Day

In Almora, the coming of broadband almost two years ago ushered a whole host of cybercafés, both government and private, from four to 20 and growing as described in the previous chapters. I visited these sites, talking to cybercafé owners and clients, checking out the history of online sites visited; the hardware and software, and the type of people frequenting these centers. The competition resembled one another. Herd-like, most of these cafés seemed to emulate each other. The centers were primarily owner-managed. These cyber-entrepreneurs came from a range of backgrounds, from retired military generals to young men trying their luck in a new line with support from their parents.

The private cybercafé can be many spaces to many people. The "after-school" center from Chapter 7 flips into a video game parlor to an official government center. Music is constantly being downloaded and listened to by both the owners and clients. But not without consequences; complaints of an owner tells of the price of leisure:

> Viruses happen a lot because people download songs from the Net to their cellphone and then come and download more out here. This takes the virus from one computer to another like a mosquito from one person to another and then we have to waste all day trying to fix this problem. It's too much work to fix!

The marriage of labor and leisure is hardly easy but inevitable. They are meant to be. Sometimes labor takes on the makings of fantasy. Becoming a "hero" is a skilled effort of photoshopping one's head to the body of "Neo" in *The Matrix* (Figure 9.1), a screen saver to remind you of your achievement. Sometimes, fantasy takes on a romantic twist, breaking national boundaries (Figure 9.2).

These cafés attribute their survival to their prime savior – Orkut. The humble presence of Orkut, a social-networking site, gives sustenance to many. A pioneer in this new game, Govind Bhisht speaks about this phenomenon:

> I am a retired military man. After the army, I decided to start a phone booth with STD/ISD calls so many tourists could use it and even the locals. Then I got my first computer some years ago but that was a dial up connection. It was too unreliable and kept getting disconnected so no one was interested. After all, how much can you do offline? You can use *Word* but for what…work purposes

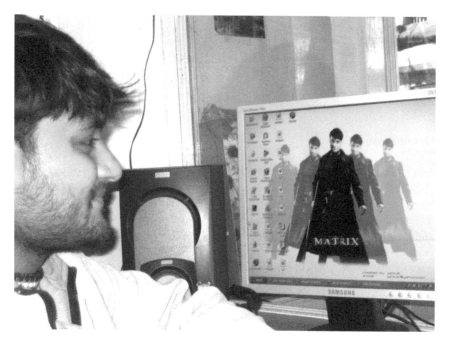

Figure 9.1 *Matrix* **Appropriated**

Figure 9.2 **Digital Romance Fantasy**

maybe. Nowadays, foreigners also come but they don't want my ISD booth, they want to talk to the computer. But honestly, it's the students that keep this place going. It's changed so much. They sit all day doing these friendship sites like Orkut. Day in and day out…if it wasn't for Orkut, my computers would not be used much.

Friendship reigns supreme. From reconnecting to sustaining relations, the youth in Almora have found a new hub to idle their time away through vigorous flirting, gossiping and chatting with one another. Particularly, teenage love finds a safe haven amidst such hardware, curtained off for privacy. With few other social spaces for "dating," many young couples come to these cybercafés, sitting in separate computer booths to chat with one another through Orkut. A unique mix of photos, hyperlinks and text, this new social arena is carefully treaded. Cybercafé staff is dismissed. Computers play cupid instead. Almost always, their friends are present at a safe distance, some interjecting strategically and others teasingly. Social yet private, cybercafés have found a new purpose. When asked why Orkut, almost all the students state the same response; "because all our friends are on it so it's good fun." Hardly profound and blatantly obvious, leisure rarely needs an excuse for its existence. It just is.

Can The Poor Come Out to Play?

We have been so busy looking at the digital divide through a utilitarian lens that little attention has been paid to how leisure is portioned out. With the top most frequented websites being social and entertainment oriented in the West (Roberts, 2006), it is easy to assume that leisure is the prerogative of the wealthier class. To test whether this common understanding applies to this context, I conducted an investigation across three intercolleges in Almora, given that youth are the prime clientele at cybercafés. With a focus on the economics of leisure, the choice of colleges are deliberately sectioned across three different economic zones, spanning urban to rural, private to public, the former being more privileged. The first site is a private intercollege in Almora town, the second is government-based, also in Almora town, and the third is a government intercollege in Hawalbagh village, almost two hours away from the town.

Having interviewed almost 100 young people from the three different intercolleges, it is evident that the private urban college has far more opportunities for access and usage of computers than the other two counterparts (see Table 9.1). The differences are stark. While youth in the private college has had a significant head start with computers since 1993, the other two colleges were provided with computers only two years ago. At home such differences persist, with 77 percent of youth from the private college owning a computer as compared to 28 percent amongst the government college in Almora town. In contrast, none of the youth from Hawalbagh intercollege have a computer at home. In the case of the government college in Almora, access to computers at "home" meant their hostels, given that the majority of them had come from far off villages to pursue higher education. Of course, access is not just about owning computers but of the time spent with these tools both at school and beyond.

Table 9.1   **Comparison of Intercolleges on Computers in Almora**

Location	Almora Town	Almora Town	Hawalbagh village
Type	govt-town	private-town	govt-village
School Grade (16–18 years)	12th Grade	12th Grade	12th Grade
Computer-Student Ratio	1:3	1:2	1:5
Total Students interviewed	25	20	52
Time per week with Computer	Twice per week – 35 min each	Everyday – 3 practicals/ 2 theory	Once a week/ 5 min per child
First installation of computers	2007	1993	2007
% used cybercafé	12	85	4
% with computers at home	28	77	0
% with cellphone	84	100	71

As noted, the students in the rural section reported spending an average of five minutes on the computer per week, contrasting drastically with students in the private college with daily sessions at school, three practical and two theoretical. This is partly attributed to the computer-student ratio of 1:5 in the government village college versus 1:2 in the private college. This evidently affects the teaching style in schools. In discussion with students in the government intercollege in Almora town, some of their difficulties come through:

Student 1: If we have to do a school project it takes so long to complete but if we have a computer at home then we can do it in only three days.	छात्र 1 : हमें स्कूल का कोई प्रोजैक्ट करना हो तो इसे पूरा करने में काफी समय लग जाता है. परंतु यदि हमारे घर में कम्प्यूटर हो तो हम इसे सिर्फ तीन दिन में कर सकते हैं।
Student 2: Also our number does not even come sometimes in the week…maybe once a week we get a chance to use the computer.	छात्र 2 : कभींदककभी सप्ताह में हमारी बारी आती ही नहीं... स्कूल में भायद हमें सप्ताह में एक बार कम्प्यूटर के प्रयोग का मौका मिलता है
Researcher: How much do teachers know about computers?	भोधकर्ता : अध्यापकों को कम्प्यूटर के बारे में कितनी जानकारी है?

Researcher: How much do teachers know about computers?

भोधकर्ता : अध्यापकों को कम्प्यूटर के बारे में कितनी जानकारी है?

Student 3: They know a lot but they are more interested in teaching one thing to all of us and because there are so many for just one computer, the teacher just shows things to us on the screen.

छात्र 3 : वे बहुत ज्यादा जानते हैं लेकिन उनकी रुचि हम सबको एक चीज़ सिखाने में रहती है और क्योंकि इतने ज्यादा लोगों के लिए सिर्फ एक ही कम्प्यूटर है, इसलिए अध्यापक सिर्फ हमें स्क्रीन पर चीज़ें दिखा देते हैं।

Student 2: They know a lot but they have to teach the curriculum to everyone. She [their teacher] gets the senior students to teach us sometimes by going to each computer or she will show us slides on *PowerPoint* and *Word*.

छात्र 2 :उनको बहुत आता है लेकिन उन्हें हर एक को पाठ्यचर्या पढ़ानी होती है। हमारी अध्यापिका कभीदककभी सीनियर छात्रों को प्रत्येक कम्प्यूटर पर जाकर हमें सिखाने को भेजती है या वह हमें पावर पॉइंट या वर्ड पर स्लाइडें दिखाती है।

There is little need to spiral any further on such inequity between these three colleges to reaffirm the economic divide affecting the digital divide, feeding into a substantive literature on this topic (Cammaerts, 2003; Compaine, 2001; Evers and Gerke, 2004; James, 2003; Mack, 2001; Marshall, 2003; Monroe, 2004; Norris, 2001). What is less explored however, is not the difference but the commonality between these seemingly disparate groups. In all three colleges, the Net is not available. This is less due to economics and more on the issue of control. Cyberspace is seen as dangerous territory, especially in formal educational institutions (it is worth recollecting the private-public discussion from Chapter 6 here). Instead, all three colleges encourage the usage of *Encarta*, an offline encyclopedia CD to complete class projects. The principal of a private intercollege in Almora town defends her decision:

> We have to regularly address in the assembly on how not to use computers. With the Net, misusing computers happens all the time. Chatting has become some kind of social status. These students feel they know more than what we can teach them and this attitude doesn't help at all. Their poor parents are taking 20,000 rupee loans to buy them computers and for what? They sit in front of those computers and chat, do Orkut, and download music. Nothing educational happens. The parents are just emotionally blackmailed while these students spend hours missing studies just to be at those cybercafés. Whatever time they get, you'll find them wasting time at those cybercafés. This is supposed to be an information device but it's not being used that way.

The mention of "Orkut" electrified the staff room. Most teachers became highly animated, cursing its presence. They accused it of monopolizing student's time, wasting away their mind, and being a bad influence; "we are trying to guard our students," they explained defensively.

Also, regardless of economic backgrounds, most students declared similar aspirations when it came to their careers. The youth voiced interest in the usual suspects of reliable professional paths such as medicine and engineering but the majority saw themselves in computer sciences, the new career choice on the block. Thereby, regardless of class positionality, the youth continue to harbor aspirations, sometimes far beyond their current status. This is markedly different from the classic Willis (1981) study on how working class youth concertedly participate in disassociating themselves from means of social mobility, reproducing their positions in society. On the other hand, the youth here, both girls and boys, even in the village intercollege, discuss plans for further education, moving to the city and getting a high paying job.

What is most interesting is their perception of computers as a tool of leisure over that of cellphones and television. This is surprising, given that computers are positioned by the media as tools of economics and mobility. After all, most students have access to cellphones and televisions despite class differences while few have computers. Yet, regardless of this current inequity in computer access and usage, common leisure perceptions persist. Youth across the board see computers as portals of entertainment. In a discussion with the government intercollege youth in Almora, much is revealed on the deep interweaving of labor, leisure and learning with new technology in "educational" and "non-educational" settings:

Researcher: So what is the difference between cellphones and computers? [Students start to talk all at once. I tell them to speak one at a time.]

भोधकर्ता : अच्छा, सैलफोन और कम्प्यूटर के बीच में क्या अंतर है?

Student 1: The computer is more entertaining than mobiles. You can get music and movies and instant message your friends and also you can do photo work.

छात्र 1 : मोबाइल की तुलना में कम्प्यूटर से ज्यादा मनोरंजन होता है। आप म्यूजिक सुन सकते हैं और सिनेमा देख सकते हैं और दोस्तों को तत्काल संदे ा भेज सकते हैं और फोटो का काम भी कर सकते हैं।

Student 2: Maam, you can download music to the phone and play games and draw and paint and there are many songs on it too that can play at the same time.

छात्र 2 : मैम, फोन में म्यूजिक डाउनलोड कर सकते हैं और गेम खेल सकते हैं और चित्र बना सकते हैं और इसमें बहुत सारे गाने भी होते हैं जो एक ही समय में बज सकते हैं।

Researcher: So all this is with or without the Net?

भोधकर्ता : अच्छा, यह सब नेट से होता है कि बिना नेट के?

Student 2: CDs maam. Online takes too long and CDs are cheap.

छात्र 2 : सीडी से मैम... ऑनलाइन में बहुत अधिक समय लगता है और सीडी सस्ती हैं।

Researcher: And what about cellphones? What do you use it for?

भोधकर्ता : और मोबाइल के बारे में? तुम इसे किसलिए प्रयोग करते हो?

Student 3: Maam, we use it but we don't have our own.

छात्र 3 : मैम, हम इसका प्रयोग करते हैं लेकिन हमारे पास अपना नहीं है।

Student 1: We can use it [cell phone] as a camera, video, games…those kinds of things.

छात्र 1 : हम इसे कैमरा, वीडियो, और , गेम्स के लिए इस्तेमाल कर सकते हैं।

Researcher: So then what is the difference between computers and cellphones?

भोधकर्ता : तो फिर कम्प्यूटर और सैलफोन में क्या अंतर हुआ?

Student 4: We can get all the knowledge from computers through *Encarta* but not as much from cellphones. And we can make projects with the help of computers but we can't do that with cellphones. We can write and talk on the cell, but we can do that with computers, too.

छात्र 4 : जैसे हम एनकार्टा के ज़रिये कम्प्यूटर पर सारी जानकारी पा सकते हैं परंतु सैलफोन से उतनी नहीं और कम्प्यूटर की सहायता से हम प्रोजैक्ट तैयार कर सकते हैं लेकिन सेलफोन से नहीं कर सकते भले ही हम सैलफोन पर लिख तथा बोल सकते हैं, लेकिन वह तो हम कम्प्यूटर से भी कर सकते हैं।

Student 5: Maam, with computers we can't click photos but we can load photos from someone else.

छात्र 5 : मैम, कम्प्यूटर से हम फोटो नहीं खींच सकते, लेकिन कहीं और से इसमें फोटो लोड कर सकते हैं

Student 3: And videos are better on computers.

छात्र 3 : कम्प्यूटर में वीडियो ज्यादा अच्छे होते हैं।

Student 5: We can keep our favorite photos stored on the computers in our files and take it out anytime we wish and paint and do other things with it.

छात्र 5 : हम अपनी पसंद की फोटो को कम्प्यूटर पर अपनी फाइल में संभाल कर रख सकते हैं और ? जब चाहें इसे निकाल सकते हैं, इसमें चित्र बना सकते हैं साथ ही अन्य काम भी कर सकते हैं।

Student 2: Maam, with computers you can do everything – knowledge and entertainment, everything.

छात्र 2 : मैम, कम्प्यूटर से सबकुछ कर सकते हैं 'दक जानकारी, मनोरंजन और सबकुछ।

Researcher: And what about television? Is it more or less entertaining than computers?

भोधकर्ता : और टेलीविजन के बारे में क्या कहोगे? क्या इससे कम्प्यूटर की तुलना में ज्यादा मनोरंजन होता है?

Student 6: TV is also entertainment but we are always dependant on the connection.

छात्र 6 : टेलीविजन से भी मनोरंजन होता है, लेकिन हमें हमे ा कनेक ान पर निर्भर रहना पड़ता है।

Student 7: Computers are most entertaining. You can save folders and hide our folders with a password so no one else can see it.

छात्र 7 : कम्प्यूटर से सबसे ज्यादा मनोरंजन होता है। आप फोल्डर्स को सेव कर सकते हैं और इसे एक पासवर्ड द्वारा गुप्त रख सकते हैं जिससे इसे कोई और न देख सके।

Student 8: TV can go off anytime but computers have battery and it saves your work.

छात्र 8 : टीवी किसी भी वक्त बंद हो सकता है, लेकिन कम्प्यूटर में बैटरी लगी होती है और यह हमारे किये हुए काम को बचा देता है।

Researcher: So then what are the problems with computers?

भोधकर्ता : अच्छा, तब कम्प्यूटरों के साथ समस्यायें क्या हैं?

Student 6: Virus madam!

छात्र 6 : वायरस मैडम।

Student 2: But you can put antivirus software on your computer.

छात्र 2 : लेकिन तुम एंटींदकवायरस अपने कम्प्यूटर में डाल सकते हो।

Student 1: Your eyes hurt!

छात्र 1 : आंखों को नुकसान होता है।

Student 3: But madam, you can put a black screen to protect your eyes.

छात्र 3 : लेकिन मैडम, अपनी आंखों को बचाने के लिए काली स्क्रीन लगा सकते हैं।

Student 7: Your hands pain too!

छात्र 7 : हाथों में भी दर्द होता है।

Student 6: Sometimes the screen blanks out and sometimes it even hangs.

छात्र 6 : कभी स्क्रीन पर कुछ नहीं आता और कभी तो यह अटक जाता है।

Researcher: So then what do you do when that happens?

भोधकर्ता : अच्छा, जब ऐसा होता है तब तुम क्या करते हो?

Student 1: Then you have to restart.

छात्र 1 : तब फिर से चालू करना होता है।

Researcher: What's the difference between using the computer at school and at home? [Students again talk at the same time; I point to specific students to talk.]

शोधकर्ता : कम्प्यूटर को घर पर इस्तेमाल करने और स्कूल में इस्तेमाल करने, इन दोनों के बीच में क्या अंतर है?

Student 8: In school we get very little time and at home we can use it whenever.

छात्र 8 : स्कूल में हमें बहुत कम वक्त मिलता है और घर पर कभी भी इस्तेमाल कर सकते हैं।

Student 4: You can do whatever you want to do when you're not in school.

छात्र 4 : जब आप स्कूल में नहीं हैं तब जो मन में आए वह कर सकते हैं।

Student 1: Maam, they won't allow us to open our song files in school.

छात्र 1 : मैम, स्कूल में वो हमें अपनी गानों की फ़ाइल नहीं खोलने देंगे।

Student 5: Teachers are always scared that we will open the wrong things so we are also scared that we will get shouted at if we open the wrong folder. While at home, we don't get shouted at.

छात्र 5 : अध्यापकों को हमे ा डर लगता है कि हम ग़लत चीज़ें खोलेंगे इसलिए हम भी डरे रहते हैं कि ग़लत फोल्डर खोलने पर चिल्लाते हैं, जबकि घर पर ऐसे में नहीं चिल्लाते।

Student 8: If we are having fun and the bell rings, then we have to end it immediately but at home we can continue.

छात्र 8 : हमें मज़ा आ रहा हो, पर घंटी बजते ही हमें इसे तत्काल बंद करना पड़ता है, लेकिन घर में हम जारी रख सकते हैं।

Students are savvy about the usage of computers and their affordances and seem to associate labor with schooling. Leisure with computers entails a complex mix of freedom and adventure. These findings prevail across the three intercolleges. It is interesting to see that while access to the Net varies between the youth, it does not block their common pursuit for entertainment through songs and movies. Come what may, with or without the Net, leisure is accomplished.

The Tale of Two People

Divisions of the town and village persist in our understanding of computer access and usage. To better understand such artifacts alongside such persistent dichotomies, this section explores narratives of two people deliberately chosen for their diametrically opposing social positions. The focus is on their personal history with the artifact, and shifts over time in their expectations, achievements, and apprehensions with this tool. The purpose is to show that in spite of obvious class differences, common grounds on computers exist. Relations between labor and leisure are drawn concerning the computer.

Vinay is a graduate in botany, now working at a non-profit, designing and editing textbooks for schools. He grew up in Almora town. Mohini is a *balwadi* or preschool teacher in a village, about six hours from Almora town. Due to a childhood accident with her foot, she has been unable to pursue high school. She is currently running for local elections for the *panchayat* office. Due to her association with an NGO in Almora town, she makes an average of two trips a year to town.

The story of Vinay:

> I am self-taught in computers. I saw my first computer in 1989. At that time I was told to use it to type but while typing, I played with it and saw what it could do. I would type Ctrl and F and see what would happen! In 1988–89, I think, when I was doing my book on botany, we had to use a manual typewriter. I remember that statistically significant calculations needed to be done so I was told by a friend to use MS-DOS. My god, it was so exciting that after graduating, I told my friend that we should teach classes instead in computers. So we did. I made a decent amount of money giving diploma workshops to people from the cantonment area. For our first class in 1991, twelve people came to learn. We did a basic package on operating computers. We even stole the Pant Institute of Advanced sciences title [laughs]. We copied them when we started. We called ourselves Pant Institute. In the day time we would teach 10th to 12th graders and we got a few housewives too. We made it [classes] "glamorous." Equally, boys and girls came. We taught housewives *Excel*, how to switch on and off computers, *Word* and that kind of thing. Then with all the competition later on, I thought I should shift to another line of work. So I somehow managed to find this NGO job where I now work with computers and edit these textbooks online. This way I am using my botany degree too in some way. You know, before Microsoft Windows 94 arrived where there were page features for visuals, I'd have to travel to Delhi to make books. Thankfully the PageMaker 95 came and then Pentium I and then Windows 97 and then Net from dialup in 98 to Pentium II. I remember email was there in 2000 but I had no purpose for it. We did not even open any sites till 2002 I think. Later, I started using computers for booking my railway tickets and to check its status but dial up takes so long and keeps the phone engaged too at the same time so we hardly used it. It was easier just going to the railway office instead. Only after 11pm you could use it but during the day, impossible! Last year John and Karen came [two foreigners volunteering at the NGO] and they set up the broadband in the office. Now I see education sites, environment sites, newspapers, *New York Times*, *Hindustan Times*. I look up health stuff like the price of medicine or check on more information about it. I like to keep checking the market value of house goods also; that is a lot of fun.

Sometimes I check my bank statements. I apply for taxes online and I have my own pin-card now. Computers are so much more fun than cellphones because cells have too small a screen. I remember that once I had made some mistake with the taxes and they said that if I did not get it corrected and send it in two days I would be paying a fine. So luckily I scanned and sent it online and saved so much money! Hey, I also have my own account on Facebook and Friendster so you should add me! Oh, and I also like Wikipedia and Google. I can spend all day with that kind of stuff. I sometimes download music online but it's so much easier to just get a CD and play on your computer. You know, I have gone through so many computers. What else? Guess what? I also heard of Second Life so decided to see what it's about. I think it's too much for the West. It's a fantasy escaping reality maybe because people are so isolated there. Why live in this life when you can live in another they must think. Here [in Almora] we're not alone. Sometimes our neighbor drops by or there is always someone to talk to so why go online for another life? Okay so it's nice that you can make some island but what's the point? Nowadays who has the time for this?

The story of Mohini:

When I was eight, I got the foot problem. Because the school was at least five kilometers from the road at that time, it was very hard for me. Now a road has been built and it's much easier. But that time, it was so hard that I was not able to go to school. So I used to just go with my mother and she would put me in charge of the other children under a tree and we would play. Fortunately a *balwadi* opened up near me so I went there instead. This gave me a chance to learn something, otherwise I would just be sitting and playing and doing not much else. Nothing would have gone into this head of mine. I then became a *balwadi* teacher myself. For a girl like me who has only studied till Grade 8 [middle school], this is god send. I started to come here for my training to Almora. I still remember how scared I was when I first came. I would not come down from my room. Four days passed and

जब मैं आठ साल की थी तब मुझे पैर की तकलीफ हुई। क्योंकि स्कूल जानांदकआना, जो कि उन दिनों सड़क कम से कम 5 किलोमीटर दूर था वह मेरे लिए बड़ा कठिन था। अब सड़क बन जाने से यह काफी आसान हो गया है। परन्तु तब यह इतना कठिन था कि मैं स्कूल जाने में समर्थ नहीं थी। इसलिए मैं केवल मां के साथ खेतों में जाती, वे मुझे अन्य बच्चों की देखभाल को छोड़ जाती, जहां हम एक पेड़ के नीचे खेला करते। सौभाग्य से मेरे घर के नज़दीक में बालवाड़ी खुल गयी इसलिए उसके बदले मैं वहां जाने लगी। इसने मुझे कुछ सीखने का अवसर दिया, नहीं तो मैं सिर्फ बैठी खेलती रहती और कुछ नहीं कर पाती। मेरे दिमाग में कुछ नहीं गया होता। फिर मैं स्वयं बालवाड़ी की ि्रिक्षिका बन गयी। मेरी जैसी केवल आठवीं कक्षा तक पढ़ी लड़की के लिए यह भगवान का वरदान जैसा है। मैं प्रि क्षण हेतु अल्मोड़ा आने लगी। मुझे अभी तक याद है कि जब मैं पहली बार आयी तब कितना डरी थी। मैं तो अपने कमरे से भी बाहर नहीं निकलती थी। चार दिनों के बाद उन्होंने मुझे नीचे आने और अन्य लोगों के साथ घुलनेंदक मिलने को मज़बूर कर दिया। मैं बहुत चिल्लाई। तब मैं अपना चेहरा भी नहीं दिखा पा रही थी, मैं इसे ढक रही थी। ऐसा लगता मानो वे सब मुझे

160 *Dot Com Mantra*

then someone forced me to come
down and be with others. I cried
so much. I couldn't even show my
face then. I would cover it. I felt
everyone would just eat me up. I
was 13 years then. When I was 18,
the NGO sent me to Delhi for my
operation. Then after that, I had
three operations. They put a rod
and said my leg would be okay
but then the rod gave me so much
fever that they removed it. Then
they did something else until the
two bones met. My one leg by
then became much shorter than
the other so they gave me a special
shoe for that leg. Now I am very
close to everyone here and I mix
with everyone. I visit close-by
villages and talk with people I've
never met before. I ask questions
now openly. I have no fear now.
If I see the *Gram Pradhan* (head
of the *panchayat*) not working, we
all get together and ask him why?
Our village that way is very nice.
We have two government schools
which are very good. Teachers
there actually show up and teach.
We know them personally from
the *balwadi* days so if they don't
come to school, we go to their
house and ask what happened.
We even got a fancy pharmacy
building in our village from the
government one time. It was so
nicely painted and all. When
someone got sick, we had to take
them on a *doli* [cot] as you know.
Once we reached the pharmacy
after climbing down the hill
with the *doli*, what did we see, a
completely empty building inside!
We were so angry. We protested
a lot. Now we have a full time
doctor there from Haldwani. He is
open even on Sundays and doesn't

खा जायेंगे। तब मैं 13 साल की थी। जब मैं
18 की थी, तब संस्था ने मुझे मेरे ऑपरे ान
के लिए दिल्ली भेजा। उसके बाद मेरे तीन
ऑपरे ान हुये। उन्होंने एक रॉड लगा दी
और कहा कि मेरी टांग ठीक हो जायेगी,
लेकिन फिर रॉड के कारण इतना ज्यादा
बुखार हुआ कि उन्होंने इसे निकाल दिया।
फिर दोनों हड्डियां जुड़ने तक उन लोगों ने
कुछ और किया। उससे मेरी एक टांग दूसरी
से छोटी हो गयी इसलिए उन्होंने मुझे उसके
लिए एक खास जूता दिया। अब मेरा यहां
हरेक से गहरा जुड़ाव है और मैं सबके साथ
घुलंदकमिल जाती हूं। मैं नज़दीक के गांवों
में जाती हूं और अपरिचित लोगों के साथ भी
बातचीत करती हूं। अब मैं खुलकर सवाल
पूछती हूं। अब मुझे कोई डर नहीं है। अगर मैं
देखती हूं कि ग्राम प्रधान काम नहीं कर रहा
है, हम सब इकट्ठे होकर उससे कारण पूछते
हैं। वैसे हमारा गांव बहुत सुंदर है। हमारे यहां
दो सरकारी स्कूल हैं जो बहुत अच्छे हैं। वहां
अध्यापक वास्तव में दिखाई देते व पढ़ाते हैं।
हम बालवाड़ी के दिनों से उन सभी को निजी
तौर पर जानते हैं, इसलिये यदि वे स्कूल नहीं
आते तो हम उनके घर जाकर पूछते हैं कि
क्या बात हो गयी है? हमारे गांव में सरकार
द्वारा महंगा औषधालय भवन भी बना था।
इसे बहुत ही बढ़िया ढंग से पेंट किया गया
था और सब कुछ। जब कोई बीमार पड़ता
तो उसे डोली में लेकर जाते थे जैसकि तुम
जानते हो। एक बार डोली सहित पहाड़ की
उतराई के बाद वहां पहुंचने पर हमने क्या
देखा औषधालय भवन अंदर से भवन बिल्कुल
खाली पडा है। हमें बहुत गुस्सा आया। हमने
बहुत विरोध किया। अब वहां हल्द्वानी से एक
पूर्णकालिक डॉक्टर आ गया है। वह इतवार
को भी खोलता है और अक्सर घर भी नहीं
जाता। इस तरह से हमारा गांव काफी सक्रिय
है। अब जूनियर स्कूल में कुछ कम्प्यूटर आ
गये हैं। लेकिन उनकी जानकारी किसी को
भी नहीं है और वहां कोई सिखाने वाला भी
नहीं है। वास्तव में, अगर मैंने इसे सबसे पहले
संस्था के दफ्तर में नहीं देखा होता तो मैं
खुद भी नहीं जान पाती कि वह क्या है। जब
मैं पहली बार यहां आयी तो मैंने इतने सारे
लोगों को इन टेलीविजनों के सामने बैठे देखा
दक तब मैंने यही सोचा। मैंने मन में कहा कि
इन लोगों के पास इतना ज्यादा वक्त है कि

even go home often. Our village that way is very active. Now some computers have come to the junior school but no one knows what they are and there is no one to teach. In fact, I would not have known what it was myself if I hadn't first seen it at the NGO office. When I came here the first time, I saw so many people sitting in front of these TVs – that's what I thought then [laughs]. I thought to myself, these people have too much time that they watch TV all day! I kept that to myself. Slowly I became close to Madhu who used to work here, so I asked her why people here watch so much TV during the day. She had a good laugh. Now I know that computers are work!... I'm not afraid of Delhi, I know my parents are. Actually for us, Nainital is like going to America, forget Delhi. But still I'm not afraid. I want to do something for my village, make it stronger. I don't want to marry ever. I just want to serve my village and my family. My parents have done so much for me, for us. My three sisters are now married and in their in-laws home. I told them to go ahead don't worry about our parents. I will take care of them, as they know I will never marry. You know, all those expectations are there on the boys. I have taken on my shoulders now for my parents – I will be the son for them. I will get a higher education, maybe learn computers someday, who knows! Oh yes, and there is this six-year-old in my *balwadi*. He sticks to me always. I ask him what he wants to become. He says, 'didi [sister], forget doctor or a computer job, that won't

ये दिन भर टीवी देखते हैं। उसे मैंने अपने तक ही रखा। धीरेंदकधीरे मेरी मधु के साथ घनिश्ठता हो गयी, जो यहीं काम करती थी। मैंने उससे पूछा कि यहां काम करने वाले दिन के वक्त इतना अधिक टीवी क्यों देखते हैं। वह खूब हंसी। अब मैं जानती हूं कि कम्प्यूटर काम के लिए है!... मुझे दिल्ली से डर नहीं लगता, मैं समझती हूं कि मेरे माता दकपिता को लगता है। वास्तव में हमारे लिए नैनीताल ही अमेरिका जाने जैसा है, दिल्ली की छोड़ो। लेकिन फिर भी मैं नहीं डरती हूं। मैं अपने गांव के लिये कुछ करना चाहती हूं, इसे मज़बूत बनाना चाहती। मैं भाादी कभी नहीं करना चाहती। मैं सिर्फ अपने गांव और परिवार की सेवा करना चाहती हूं। मेरे मातांदकपिता ने मेरे, हमारे लिए इतना किया है। मेरी तीन बहिनें अब विवाहित हैं और अपने ससुराल में हैं। मैंने उनसे कहा कि मातांदकपिता की चिंता छोड़ आगे बढ़ो। मैं उनकी देखभाल करूंगी जैसा कि वे जानते हैं कि मैं कभी भाादी नहीं करूंगी। आपको पता है, सारी उम्मीदें लड़कों से रहती हैं। अपने मातांदकपिता की जिम्मेदारी मैंने अपने कंधों पर ली है। मैं उनके लिए लड़का होऊंगी। मैं ऊंची पढ़ाई करूंगी, कौन जानता है भाायद किसी दिन कम्प्यूटरों के बारे में सीखूं। हां, बालवाड़ी में ये 6 साल की उम्र का बच्चा है। वह हमे ाा मुझसे चिपका रहता है। मैं उससे पूछती हूं कि वह क्या बनना चाहता है। वह कहता दीदी डॉक्टर या कम्प्यूटर के काम को छोड़ो, वह नहीं होगा लेकिन बड़ा होकर मैं फौज में भर्ती हो सकता हूं, तब मैं कुछ बन जाऊंगा... देखती हूं, यदि कर पायी तो मैं और ज्यादा पढ़ूंगी और एक दिन स्मार्ट बन जाऊंगी। भाीघ्र ही मैं इण्टर कॉलेज में जाऊंगी और दूसरी लड़कियों को भी इसके लिये तैयार करूंगी। मैं इसे पक्का करूंगी कि अन्य लड़कियां भी अपने गांव के लिए कुछ करें, कुछ उपयोगी जैसे साइंस या कम्प्यूटर, कुछ भी जो हमारे गांव का विकास करे। देखो, मैंने आपको बताया कि मैं कितनी मूर्ख थी कि कम्प्यूटरों को टीवी समझ रही थी। मैं चाहती हूं कि बच्चे जब पहली या दूसरी कक्षा में हों तभी से वे कम्प्यूटर भाुरू कर दें इससे वे आरंभ से ही सब कुछ जानेंगे; तभी वे ज़िंदगी में आगे बढ़ेंगे, नहीं? देखिये, यहां धीरेंदकध ीिरे जींस, अंग्रेजी और अब कम्प्यूटर आ रहे हैं; धीरेंदकधीरे हमारी बेहतरी व तरक्की हो

happen but I can join the army when I grow up, then I'll become somebody'…See, if I can, I will study more and someday become smart. I will soon go to intercollege and I will make other girls join too. I will make sure these girls do something too for their village, something useful like science or computers, anything to grow our village. Look, I told you how silly I was to mistake the computers for TVs. I want children to start computers when they are in the first or second grade so they will know everything from the start, then only they can move in life no? Look, slowly now jeans, English and now computers are coming. We are slowly becoming better and moving ahead here. I sometimes joke when people keep asking me what happened to my foot and why do I wear one big shoe and one small shoe. I tell them it's the fashion these days in my village and then I tell them don't worry, it will come soon to yours as well [laughs]. Didi, what do you do in your village?

रही है… मेरे पैर को क्या हुआ, मैं एक पैर में बड़ा और दूसरे में छोटा जूता क्यों पहनती हूंदक जब लोग मुझसे ये सवाल पूछते हैं तो मैं कभीदककभी मजाक में कहती हूं कि यह आजकल मेरे गांव में इसी का फै न है, और फिर कहती हूं, चिंता मत करो जल्दी ही यह तुम्हारे यहां भी आने वाला है (हंसती है)। दीदी, आप अपने गांव में क्या करती हो?

The Blasphemy of Leisure

The days of leisure have arrived online. The "privileged" it is seen, prod, chat, flirt, pry, share, and play. Congregations of people online express thoughts from the profound to the mundane. So what are we to make of a similar social frenzy in context of the supposed developing world? How do we comprehend those that surf the electronic "flow," not for gaining empowerment per say but for the pure fun of it? Is it blasphemous to say that the digital "have-nots" can and will play amidst their plight of poverty, isolation and deprivation, "wasting away" this cost-intensive opportunity to labor? Can we justify US 3 billion dollars in investments in computers for "mere" entertainment? Children in rural India, South Africa, and Egypt are witnessed chasing each other online in car games; couples romancing online, youth downloading music and videos, Skype and email. In my own fieldwork experiences in rural Andhra Pradesh, I document how village youth take to video games and movies over pragmatic services such as ration cards, soil testing, crop pricing, and health information. Entertainment is seen to take precedence over economic-oriented tasks in rural arenas (Arora, 2005, 2006b, 2008a).

In a recent study of rural computer kiosks in India, it was observed that, "even the poorest populations have desires that go beyond those required for physical sustenance" (Rangaswamy and Toyoma, 2006, p. 3). Mass media has been and continues to be a vital force in rural India. Village folk are neck to neck with their urban counterpart when it comes to entertainment – popular soap operas, television serials, and music, leveraging on a range of old and new technologies from the radio, cable, cellphones and the Net. Recreation is at the heart of village life, extended by new technologies:

> From field ethnography, we find that urban youth slang and speech styles do not lag behind in villages. Neither do communication styles and channels. Instant messaging is immediately embraced by younger kiosk operators. Fan clubs of matinee idols bring in youth fashion and trends along with film music. Most popular films and film music are released within a month in hub-towns. Cassettes, pulp-film magazines, and even VCDs are snapped up quickly by rural consumers. We found in one case, that women from a village in Tamil Nadu flocked to a rural kiosk where an online celebrity chat was organized with the director of a contemporary soap opera. (p. 5)

Even with a shortage of money, villagers collaborate to gain access to entertainment through shared cable expenses. One-third of cable TV installations in India happen to be in rural areas (Cooper-Chen, 2005). Strong value is placed on entertainment even as people in poor areas continue to struggle for their basics. Contradictory to Maslow's (1954) seminal theorization on human motivation, the predictive hierarchy of needs is disbanded as entertainment oversteps physiological wants. In fact, Miller and Slater (2000), in their pioneering study of Net usage amongst Trinidadians, warns us to not get seduced with the altruistic notion of initiating and "domesticating" Third-World nations with new technology. They claim that such communities are already attuned and completely engaged with computers through online gossip and "saucy" public flirting. So, perhaps we need to make the case that the computer as a tool of empowerment may be getting retooled for "less noble" purposes as that of pleasure and leisure.

Russell (1960) reminds us that leisure has long been contested along class lines. So, it is not outrageous to say that this continues to date as people in prosperous countries are expected to "hang out" online while the rest of the supposed Third-World nations are meant to log on so they can diligently and laboriously climb up the socio-economic ladder through "necessary" government mandated employment and health information sites.

There is an assumption that the poor will somehow behave differently from their wealthier counterpart. Herein lays a deep-seated double bind of contemporary ICT development thinking. The neoliberal view espouses that the poor will "leapfrog" conventional and chronic barriers for higher socio-economic mobility. Yet, if equity between the Third and First-World is to be achieved, we should expect that the poor, just as the rich, the rural, just as the urban folk, will use computers for "frivolous" and "trivial" purposes. One can argue that this persisting tension stems from a morality of poverty where the pragmatic and ameliorative are the main benchmarks concerning Third-World computing. And through this narrowed lens, we can miss the actual engagements and ingenious strategies that the poor employ to cope and escape from their current plight. Entertainment is a key tool here with class taking a backseat.

Yet, the government continues to produce and promote "necessary" knowledge (Calhoun, 2002), prompted by the advent of computers. It is the nature of the beast. Governments strive to be the "ultimate custodians of social order" (Roberts, 2006, p. 12), where computer engagements and information consumption become part of this expansive territory. Thereby, it is natural for the State to be acutely interested in all kinds of phenomenon of citizen action, including that of leisure. After all, leisure is not necessarily harmless or virtuous. From pornography to political blogging, what starts as leisure, can take on more serious consequences. In fact, other actors are just as interested, if not more, in leisure acts with computers: IT corporations need to sell and design interfaces for the new consumer base, entrepreneurs need to tailor online services for profit-making and development agencies and schools focus on usage for educational engagement.

A larger problem, however, has to do with the positioning of labor and leisure at opposing ends of the development spectrum. Old class theories demarcated these two realms where work and play were bounded and separate from one another. The modern division of labor views leisure as that which needs to be earned. In the recent decade or so, the shift has been from dichotomy to dialecticism. The organization and perception of work has undergone change. Compartmentalized and rationalized thinking of these two realms have given way to a sophisticated intermingling of play and labor. As we have seen in prior chapters, play is not easy and there is much labor embedded in good play.

In fact, "serious leisure" (Stebbins, 1992) can provide long-term accomplishments and deep rooted skills through gratification. Besides, leisure is deeply educative:

> People can develop skills and discover abilities that would otherwise have been untapped. Once again, this is possible because in leisure people can experiment and take risks without failure having devastating consequences. Through leisure activities, children and adults can develop motor, language and social skills, which may then be transported into other areas of their lives. Play methods work in classrooms and in many other places as well. (Roberts, 2006, p. 7)

The benefit in paying attention to leisure with computers is in their potential social effect of binding people and contributing to personal health, well-being and fulfillment through sustenance of relationships and overall life satisfaction. Of course, the "harmful" effects of such pursuits tend to gain more attention, given its economic and social ramifications, such as the industry of porn. Regardless, the point is not to debate the virtue of leisure. Instead, when concerning the field of development, we should start to take seriously the other side of the much focused upon labor – that of leisure. After all, as Roberts astutely argues, "the different classes do not do different things so much as more and less of the same things" (2006, p. 66). Thereby, the *Right to Labor* goes hand in hand with the *Right to Play* and, in doing so, equity in leisure achieves center stage.

Chapter 10

Conclusion

This book addresses contemporary understandings, initiatives and concerns of social development with computers, particularly in Almora, a small town in Uttarakhand, India. The story of the field of development is retold to remind the reader of the continuities of such initiatives and the special place technology occupies in visions of building nationhood, specifically in India. The argument is made that Gandhian self-sufficiency with local and "appropriate" technologies makes its way to contemporary propositions of empowerment with computers, but not without controversy. The popular notion of "leapfrogging" socio-cultural barriers is addressed, given its underlying and fundamental influence on IT policy for development in India. This goes hand in hand with new expectations and approaches to people and artifacts, particularly the computer, in facilitating economic and social empowerment.

The choice of ethnography as a methodology to investigate new technology has been fundamental in unpacking popular beliefs and conceptions of computer access and usage. An important ingredient is context, with an emphasis on informal social spaces, to understand interaction between actors and artifacts. In doing so, this text emphasizes the dearth of open investigations into the access and usage of computers in remote and marginalized areas without an ameliorative angle or causal outlook. Thereby, to best understand social learning *about* and *with* computers amongst a relatively new, and for the most part, historically marginalized people in Almora, this work reveal the conditions of their situatedness. Of course, this book is not a recipe to recreate Almora in its entirety, an impossibility at large. Rather, it's the making known of a complex and perhaps "exotic" culture to the reader – to discover familiarity, and fragments of social worlds, within which this work is positioned.

The objective here is to strategically amalgamate Almora's "rurality" and "urbanity" as well as its "local" and "global" dimension. In other words, the attempt here is to capture some of the flows of people and events that constitute the makings of a computing practice. In which case, Almora becomes a place of *swamis* and *sadhus*, a historic site of rural development, feminism and environmental activism, and a utopia for international eccentrics and seekers at large; a retirement community for military officers, a summer escape for the Delhi elite, a breeding ground for NGOs, and ideologies of new Statehood and technology for social change. Here, Bob Dylan meets British pilots of the Second World War that meet village girls sporting lipstick and high heels. A bias is revealed for "novelty," privileged over the much documented "conventionality" of village archetypes.

Overall, this manufactured immersion hopes to shift the status of the reader from voyeur to spectator to an invested social actor of a less exotic terrain.

Questions beget questions: What constitutes the history of a technology? Are technologies neutral? Do new technologies produce new practices? Are new artifacts necessitated by the limitations of the old? In other words, does new mean better? And if so, in an aspiring equitable society, should not all be on the right side of the tracks or divide as you may, making redundant such divisions? And what are we to make of different social practices with a new technology across different cultures? Is this a temporal lag in which initial adaptation will eventually become standardized practice? Narratives in this text serve to make concrete such pontifications that lead to further investigation.

This text reveals Almora to be less of a byproduct of a linear trajectory of chronological events and more of a hybridization of people, processes, and products, transnational and cross-cultural. Here, computers are part of this larger construct and context, as an artifact as well as a technique. This approach shows that new technology can produce age-old social practices just as old technology can birth new social acts. If we are to discount popular conceptions of technology as predictive, how do we go about looking at new technology usage? Shall we just demote the artifact as just "another tool" or position it as a legitimate actor in a complex orchestration of people, places and things? How much of this interactivity needs to be situated in its historicity and how much of that past can be seen to continue to the present? How enduring are these relationships and actors? While analyzing new technology usage in isolation may make for a dramatic emphasis, this text takes the stand that, by engaging in the "messiness" of human encounters and learning, rituals both old and new, much can be understood of the politics of a new technology in Almora and beyond. For instance, as we see in Chapter 4, the usage of the cell phone in maintaining social networks hints at the continuing and strengthening migration of males to the city and the geographical fragmentations of the family unit. Or, for example, the intervention of the water pump reveals the unwanted effect of gender isolation, wherein bonding practices amongst new brides through communal fetching of water ceases as access to water becomes convenient and individualistic. The reader also gets a glimpse at how government development schemes and actors can produce a system of cynicism and distrust that flows into "new" schemes. In other words, the focus is not just *on* "things;" it's an emphasis on *doing* things.

Throughout this text, we encounter the discussion of "intermediaries." The computer's supposed neutrality and objectivity is pitted against human foibles. We see that in Chapter 5, as new and heavily invested public-private sector schemes in agriculture strive to provide farmers with access to computers to circumvent middlemen. Yet, this investigation demonstrates that, far from computers being the new intermediary of "essential" and "relevant" agricultural information, they become continuations of existing "experts." As with middlemen, computers are just one factor in the struggle as farmers navigate through the maze of institutions and policies that continue to pose as formidable barriers to socio-economic mobility.

What is emphasized is the tremendous knowledge that farmers already possess ranging from e-waste, kiwi production, and market politics to the complexities in the relationship between education and employment. In other words, far from the "scarcity" paradigm that frames State policy, there seems to be a wide and diverse range of knowledge amongst these farmers. If anything, scarcity is less in the arena of information and more in life choices, from education to employment. Besides, the nature of "knowledge" that farmers possesses is not solely about "farming," reminding us to not reify entire populations and their knowledge through a particular lens. Here, nuanced understandings of knowledge amongst farmers are revealed, where relations with State, corporate and local "experts" are far more entwined than opposing. The chief argument is not to prove a triumph of "local" over "standardized" knowledge or reveal subjugations and exploitations, but to demonstrate that decision-making is not primarily based on access to "relevant" information. Instead, it's a dynamic and complex process that is contingent upon temporal constraints and opportunities as well as certain desires and perceptions that are seen as factual.

This text underlines that intermediaries of institutions and individuals in knowledge formation are here to stay, in spite of the computer as a possible new intermediary. The teacher as intermediary (or lack thereof) in Chapter 6, reinforces this argument. Being "free" to learn doesn't necessarily translate to being "free" from intermediaries. In fact, if used well, intermediaries of persons and institutions such as schools *can* be key actors to more freedom. Thereby, the politics of learning with computers are all about navigating constraints and opportunities. By accepting some of the constraints of formal schooling, computers may be protected within its informal spaces. The risk is in degree and not kind. In fact, even with the minimal presence of teachers, they can be decisive factors in the interactivity with computers. That said, it is important to remind ourselves that "Hole-in-the-Wall" (HiWEL) is not just an experiment with education and technology but a romance with an idea that got the world's attention, that of *free* learning – free from schools and their overarching curriculum. In this flurry of excitement, key questions get lost: is collaborative learning equivalent to democratic learning? Are playgrounds and other such "free" informal spaces less constrained than formal spaces and more importantly, are they more conducive to learning? Is informal learning necessarily better? Can informal and formal practices co-exist and if so, how?

As we see in the private cybercafé in Part III, the staff including myself, makes all attempts to serve as extensions of the artifact, many times suppressing our own intent. At times, however, deliberate intervention takes place as seen in Chapter 7 when the owner and staff collude to create the biology assignment for the student "client" through specific and detailed instructions. Their conscious roles as intermediaries lead to "success" in accomplishing the task. Thereby, the role and impact of intermediaries are subject to their context. Further, Chapter 7 deals with a highly contentious topic in academia, accelerated by the advent of computers and the Net – that of plagiarism: of cutting and pasting texts to create a thesis, hunting material to make it one's own, and conspiring with others to conceal ownership and authorship. However, the term plagiarism conjures notions of immorality and intentionality, lack of originality and learning that can serve as a barricade to fresh thinking on this issue. Instead, much is learnt here by focusing on the range and kinds of learning that goes into the harnessing of multiple resources at hand, digital and otherwise, to accomplish schooling tasks. Issues of direct versus indirect interfacing with computers, digital learning, the private-public nature of cybercafés, their relationship with schools, gender and collaborative learning with computers and educational content are explored. The intent is to sideline the "immorality" of plagiarism, and instead look into the actual nature and implications of learning outside formal educational institutions through computers.

We do not just look at who uses computers and through whom but also, for what purposes. The euphoria on digital freedom and choice has already been humbled to some degree in Chapter 5. But within every myth is ingrained some truth. The current frothing on divisions between "info-rich" and "info-poor" has drawn attention to the imbalance in content access across cultures and people. But if we are to accept this divide, we would have to believe in *a* body of knowledge as sacrosanct and autonomous, an essential fuel to the deprived minds of the supposed Third-World countries. So be it "essential" information for students to farmers, there is an implicit heeding to the notion of opening doors to a self-contained box of treasured knowledge. However, this boxing of knowledge can be a Pandora's Box. Far from virtuosity, we see information floating in disjunctured spaces, amongst false concepts, perpetuating mis-information. For instance, Chapter 8 focuses on the shopping spree online for information, giving us the Westernization of Native American art, the beautification of the classic Ravi Varma and the uncomfortable positioning of Mona Lisa with bathroom tiles. Mockery to some is serious work to another. Such concoctions of information online is shaped into a fact and bounded as curriculum by the students. Learning, after all, is an act of consumption. As the educational bazaar is seen to expand from local to global, particularly with the help of new technology, there is a belief that learning will make that shift too. There is hope that young participants of this information age will take charge, and push boundaries of knowledge to new frontiers. The Net is their global curriculum. However, these beliefs get challenged through these learning events at the cybercafé. The girls' interactions with art images reveal their criteria for what qualifies as "Western" and "Indian." In this process, we delve into aspects of information and knowledge, consumption and production of knowledge, and globalization of learning through the Net. Here, interaction does not necessarily equate with understanding, learning engagements with new technologies can be peripheral and fleeting and that which gets learnt can diverge far from what is expected to be learnt.

In fact, tremendous effort goes into the making of "facts." Information is dialectically constructed as in the case of the farmers and students. Truth and falsity of information take a backseat to the actual makings of content. Extensive labor is required to shape information through continued discussions, navigations, weighing of options, strategic positioning, and presentation. Even in entertainment as seen in Chapter 9, much is learnt on downloading, viruses and more. In the marriage of labor and leisure, learning is conceived. Lastly, in Chapter 9, we are reminded that international development is serious business. Stakes of poverty are high. The "helping hand" seeks to part way for a better life, a keener future, and more opportunities to labor. So, it is no coincidence that computers are seen as tools of labor and mobility in the field of education, healthcare, and employment amongst policy makers and media in poor nations. The higher the stakes, the more the work expected. The ingenuity of students in cybercafés is captured in how they go about laboriously completing their school projects, ascending the educational ladder.

However, we learn that these academic pursuits at cybercafés are primarily around exam time, with the regular activity overridden by entertainment. In fact, it is discovered that cybercafés' yearly survival is contingent upon the usage of such spaces for leisure purposes. Here, Orkut is king. Bollywood rubs shoulders with videogames. Horoscopes and photo galleries help "kill" time. These findings lead to an investigation on who are being entertained in such spaces, delving into the familiar world of digital equity. The revelation of differences in access to computers and the Net amongst students from public and private colleges, rural and urban does not surprise and, in fact, feeds into an already fattened body of literature on the digital divide. However, through such an activity, a surprise lies in the waiting. Despite economic differences and access to hardware and software between these groups, leisure is important to all. Computers are seen as prime portals of entertainment over cellphones and televisions. We witness diverse strategies and knowledge that students have of such artifacts and their ability to circumvent digital obstacles to achieve entertainment. Further, regardless of their economic background and current reality, they share similar career aspirations in computer science and other "prestigious" paths. The complexity of perception is pushed further through the comparative narrative of two people, a young village girl teacher with that of a well-educated man from Almora town. These tales exemplify the intricate weavings of labor, leisure and learning, of perception and reality, and urbanity and rurality. Utilitarian notions of development are challenged, gently reminding us that poor people, like "us," and perhaps more so, need to come out and play, and play they do!

Overall, this study highlights some key issues on computing such as direct interfacing and empowerment, online information and socio-economic mobility, the makings of knowledge and relations with new technology, the ameliorative role of computers and, the role of intermediaries in technology design, access and usage. This work shifts emphasis from deficiencies to constraints and opportunities. Rather than viewing computer access and usage through the much exhausted lens of State institution failure, this study looks outside of formal computing arenas and into common public spaces to better understand learning with such tools.

To summarize, computers are here to stay. Their strength lies in their vulnerability. We see how this artifact can become many things to many people: a technique, a form, a symbol, a lifestyle. In such a metamorphosis, we capture tensions between promises of socio-economic mobility and empowerment against quotidian events, trivial to the profane. Irreverence towards this new tool is underlined through the witnessing of computer usage for entertainment, plagiarism, and secret romance. Strangely enough, the most potent tool of contemporary labor in this so called developing country, India, is viewed by her youth as a portal of leisure, beating strong competitors such as the cell phone and television. This is not to say that leisure has supplanted labor in its association with this artifact. Instead, it is argued that computers need to be reconstituted to amalgamate labor and leisure. After all, the promise of modernity, globalization, prosperity and wealth through new technology is too alluring to be dismissed easily. Elasticity of perceptions saves the day.

Bibliography

Agarwal, B. (1992). The gender and environment debate: Lessons from India. *Feminist Studies*, 18(1), pp. 119–58.

Aggarwal, D.D. (2002). *History and development of elementary education in India*. New Delhi: Sarup and Sons.

Agrawal, A. (2005). *Environmentality: Technologies of government and the making of subjects*. Durham, NC: Duke University Press.

Agrawal, D.P., Shah, M., and Jamal, S. (2007). *Traditional knowledge systems and archaeology: With special reference to Uttarakhand*. New Delhi: Aryan Books International.

Agrawal, R. (2002). *Small farms, women, and traditional knowledge: Experiences from Kumaon hills*. Paper presented at the 17th Symposium of the International Farming Systems, November 17–20, 2002, Orlando, Florida.

Aikman, S. (1999). *Intercultural education and literacy: An ethnographic study of indigenous knowledge and learning in the Peruvian Amazon*. Amsterdam: Benjamins Publications.

Altbach, P.G. (1989). *Twisted roots: The Western impact on Asian higher education*. Dordrecht: Springer.

Altbach, P.G. (1993). *The dilemma of change in Indian higher education*. Dordrecht: Springer.

Altbach, P.G. and Kelly, G.P. (1988). *Textbooks in the third world: Policy, content, and context*. New York, NY: Garland Publishing.

Anderson, B. (2006). *Imagined communities: Reflections on the origin and spread of nationalism* (Rev. ed.). London: Verso.

Anderson, J. (1998). *Plagiarism, copyright violation, and other thefts of intellectual property*. Jefferson, NC: McFarland & Co.

Annamalai, K. and Rao, S. (2003). *What works: ITC's e-choupal and profitable rural transformation: Web-based information*. Washington, DC: World Resources Institute.

Appadurai, A. (1988). *The social life of things: Commodities in cultural perspective* (1st pbk ed.). Cambridge: Cambridge University Press.

Appadurai, A. (1996). *Modernity at large: Cultural dimensions of globalization*. Minneapolis, MN: University of Minnesota Press.

Arnove, R.F., Altbach, P.G., and Kelly, G.P. (1992). *Emergent issues in education: Comparative perspectives*. Albany, NY: State University of New York Press.

Arora, P. (2005). Profiting from empowerment? Investigating dissemination avenues for educational technology content within an emerging market solutions project. *International Journal of Education and Development using Information and Communication Technology*, 1(4), pp. 18–29.

Arora, P. (2006a). *E-karaoke for gender empowerment*. Paper presented at the Information and Communication Technologies for Development, May 25–26, 2006, Berkeley, California.

Arora, P. (2006b). Karaoke for social and cultural change. *Information, Communication and Ethics in Society*, 4(3), pp. 121–30.

Arora, P. (2006c). The ICT laboratory: An analysis of computers in public high schools in rural India. *Association of Advancement in Computing in Education*, 15(1), pp. 57–72.

Arora, P. (2007). Evaluating asynchronous online engagement on international security. *Electronic Journal of e-Learning*, 6(1), pp. 1–10.

Arora, P. (2008a). Instant messaging Shiva, cybercops, *Bil Klinton* and more: Children's narratives from rural India. *International Journal of Cultural Studies*, 11(1), pp. 69–86.

Arora, P. (2008b). Perspectives of schooling through karaoke: A metaphorical analysis. *Education Philosophy and Theory*, 40(3), pp. 1–21.

Arora, P. (2009). *Siliconizing youth in Indian education policy: Rearticulating "technological youth" as common sense*. Paper proceedings, Global Communication Association (GWA) Conference November 26–27, 2009, Bangalore, India.

Arora, P. (2010). Learning in the age of information. In S. Marshall and W. Kinuthia (eds) *Cases 'n' places: Global cases in educational and performance technology*. Charlotte, NC: Information Age Publishing.

Ascough, J.C., Hoag, D.L., Frasier, W., and McMaster, G.S. (1999). Computer use in agriculture: An analysis of great plains producers. *Computers and Electronics in Agriculture*, 23(3), pp. 189–204.

Baker, S. (1997). *The politics of sustainable development: Theory, policy and practice within the European Union*. London: Routledge.

Bakhtin, M.M. and Holquist, M. (1981). *The dialogic imagination: Four essays*. Austin, TX: University of Texas Press.

Banerjee, A., Cole, S., Duflo, E., and Linden, L. (2003). *Improving the quality of education in India: Evidence from three randomized experiments*. Cambridge, MA: MIT Press.

Barrow, O. and Jennings, M. (2001). *The charitable impulse: NGOs and development in East and north-East Africa*. Oxford: James Currey Publishers.

Barthes, R. (1975). *The pleasure of the text* (1st American ed.). New York, NY: Hill and Wang.

Barton, D., Hamilton, M., and Ivanic, R. (2000). *Situated literacies: Reading and writing in context*. London: Routledge.

Benjamin, W. and Tiedemann, R. (1999). *The arcades project*. Cambridge, MA: The Belknap Press.

Best, M.L. and Maier, S.G. (2007). Gender, culture and ICT use in rural south India. *Gender, Technology and Development*, 11(2), pp. 137–55.

Bhabha, H.K. (1990). *Nation and narration*. London: Routledge.

Bhatnagar, S.C. and Schware, R. (2000). *Information and communication technology in development: Cases from India*. Thousand Oaks, CA: Sage Publications.

Bhatt, K.N. (1997). *Uttarakhand: Ecology, economy, and society*. Allahabad: Horizon Publishers.

Blewitt, J. (2008). *Understanding sustainable development*. London: Earthscan.

Bourdieu, P. (1984). *Distinction: A social critique of the judgement of taste*. Cambridge, MA: Harvard University Press.

Brook, J. and Boal, I.A. (1995). *Resisting the virtual life: The culture and politics of information*. San Francisco, CA: City Lights Books.

Bugeja, M. (2004). Cellphones and real-world communication. *Education Digest*, 70(3), pp. 36–9.

Buranen, L. and Roy, A.M. (1999). *Perspectives on plagiarism and intellectual property in a postmodern world*. Albany, NY: State University of New York Press.

Burbules, N. (2000). Aporias, webs and passages: Doubt as an opportunity to learn. *Curriculum Inquiry*, 30(2), pp. 171–87.

Butler, C.W. (2008). *Talk and social interaction in the playground*. Aldershot: Ashgate.

Calhoun, C. (2002). Imagining solidarity: Cosmopolitanism, constitutional patriotism, and the public sphere. *Public Culture*, 14(1), pp. 147–71.

Cammaerts, B. (2003). *Beyond the digital divide: Reducing exclusion, fostering inclusion*. Brussels: VUB Brussels University Press.

Canagarajah, A.S. (1999). *Resisting linguistic imperialism in English teaching*. Oxford: Oxford University Press.

Castells, M. (1996). *The rise of the network society*. Cambridge, MA: Blackwell Publishers.

Castells, M. (2000). *End of millennium* (2nd ed.). Malden, MA: Blackwell Publishers.

Certeau, M. de (1984). *The practice of everyday life*. Berkeley, CA: University of California Press.

Chambers, R. (1983). *Rural development: Putting the last first*. New York, NY: Longman.

Cheater, A.P. (1999). *The anthropology of power: Empowerment and disempowerment in changing structures*. London: Routledge.

Chu, H. (2008). Where every inch counts. *Los Angeles Times*, September 8.

Chudacoff, H.P. (2007). *Children at play: An American history*. New York, NY: New York University Press.

Cleaver, F. (1999). Paradoxes of participation: Questioning participatory approaches to development. *Journal of International Development*, 11(4), pp. 597–612.

Cockburn, C. (1991). *Brothers: Male dominance and technological change* (New ed.). London: Pluto.

Cockburn, C. and Furst Dilic, R. (1994). *Bringing technology home: Gender and technology in a changing Europe.* Buckingham: Open University Press.

Cockburn, C. and Ormrod, S. (1993). *Gender and technology in the making.* London: Sage Publications.

Cohen, K.R. (2005). *Who we talk about when we talk about users.* Paper presented at the Ethnographic Praxis in Industry Conference, November 14–15, 2005, Redmond, WA.

Compaine, B.M. (2001). *The digital divide: Facing a crisis or creating a myth?* Cambridge, MA: MIT Press.

Cooper-Chen, A. (2005). *Global entertainment media: Content, audiences, issues.* Mahwah, NJ: Erlbaum.

Cozic, C.P. (1996). *The information highway.* London: Thomson Gale.

Dangwal, R., Jha, S., and Kapur, P. (2006). Impact of minimally invasive education on children: An Indian perspective. *British Journal of Educational Technology,* 37(2), pp. 295–8.

Das, G. (2000). *India unbound.* New Delhi: Viking.

Deleuze, G. and Guattari, F. (1976). *Rhizome: Introduction.* Paris: Éditions de Minuit.

Dewey, J. (1916). *Democracy and education.* Chicago, IL: Southern Illinois University Press.

Diamond, J.M. (2005). *Guns, germs, and steel: The fates of human societies.* New York, NY: Norton.

Diamond, J.M. (2006). *Collapse: How societies choose to fail or succeed.* New York, NY: Penguin.

Dickey, S. (2000). Permeable homes: Domestic service, household space, and the vulnerability of class boundaries in urban India. *American Ethnologist,* 27(2), pp. 462–89.

DIT (2005). *Draft framework for establishment of 100,000 common service centres.* Department of Information Technology document, 1(1)/2005 EGD, New Delhi, India.

Donner, J. (2008). Research approaches to mobile use in the developing world: A review of the literature. *The Information Society,* 24(3), pp. 140–59.

Donner, J., Gandhi, R., Javid, P., Medhi, I., Ratan, A., Toyama, K. and Veeraraghavan, R. (2008). Stages of design in technology for global development. *Computer,* 41(6), pp. 34–41.

Doyle, D., Jolly, R., Hornbaker, R., Cross, T., King, R.P., Koller, E.F., Lazarus, W.F. and Yeboah, A. (2003). Case studies of farmers' use of information systems. *Review of Agricultural Economics,* 22(2), pp. 566–85.

Drèze, J. and Sen, A.K. (2002). *India, development and participation* (2nd ed.). Oxford: Oxford University Press.

Drucker, P.F. (1995). *Managing in a time of great change.* New York, NY: Truman Talley Books.

Duranti, A. and Goodwin, C. (1992). *Rethinking context: Language as an interactive phenomenon.* Cambridge: Cambridge University Press.

The Economist (2008, March 8). Battling the babu raj – India's civil service (Indian administrative service).

Eisner, C. and Vicinus, M. (2008). *Originality, imitation, and plagiarism: Teaching writing in the digital age*. Ann Arbor, MI: University of Michigan Press.

Escobar, A., Hess, D., Licha, I., Sibley, W., Strathern, M., and Sutz, J. (1994). Welcome to cyberia: Notes on the anthropology of cyberculture [and comments and reply]. *Current Anthropology*, 35(3), pp. 211–31.

Evers, H.-D., and Gerke, S. (2004). *Closing the digital divide*. Sweden: Lund University, Centre for East and South-East Asian Studies.

Farooqui, A. (1997). *Colonial forest policy in Uttarakhand, 1890-1928* (1st ed.). New Delhi: Kitab Publishing House.

Featherstone, M., Lash, S., and Robertson, R. (1995). *Global modernities*. London: Sage Publications.

Fisher, W.F. (1997). Doing good? The politics and anti-politics of NGO practices. *Annual Review of Anthropology*, 26(1), pp. 439–64.

Fiske, J. (1991). *Understanding popular culture*. London; New York: Routledge.

Foucault, M. (1977). *Discipline and punish : The birth of the prison* (1st American ed.). New York, NY: Pantheon Books.

Frank, A.G., Chew, S.C. and Denemark, R.A. (1996). *The underdevelopment of development: Essays in honor of Andre Gunder Frank*. Thousand Oaks, CA: Sage Publications.

Freire, P. (1986). *Pedagogy of the oppressed*. New York, NY: Continuum.

Fuller, B. (1989). Third world school quality. *Educational Researcher*, 18(2), pp. 12–19.

Gadgil, M. and Guha, R. (1993). *This fissured land: An ecological history of India* (1st University of California Press ed.). Berkeley, CA: University of California Press.

Gandhi, M. (1971). *Collected works*. Hyderabad: The Publications Division.

Gaonkar, D.P. (2001). *Alternative modernities*. Durham, NC: Duke University Press.

Garai, A. and Shadrach, B. (2006). *Taking ICT to every Indian village*. New Delhi: One World South Asia.

Gardner, H. (1999). *Intelligence reframed: Multiple intelligences for the 21st century*. New York, NY: Basic Books.

Gee, J.P. (1996). *Social linguistics and literacies: Ideology in discourses* (2nd ed.). London: Taylor and Francis.

Gee, J.P. (2007). *What video games have to teach us about learning and literacy* (Rev. and updated ed.). New York, NY: Palgrave Macmillan.

Geertz, C. (1973). *The interpretation of cultures: Selected essays*. New York, NY: Basic Books.

Giddens, A. (2000). *Runaway world: How globalization is reshaping our lives*. New York, NY: Routledge.

Gill, K.S. (1996). *Information society: New media, ethics, and postmodernism*. London: Springer.

Gilster, P. (1997). *Digital literacy*. New York, NY: Wiley Computer Publishing.

Ginsburg, F.D. (2002). *Media worlds: Anthropology on new terrain*. Berkeley, CA: University of California Press.

Gitlin, T. (2001). *Media unlimited: How the torrent of images and sounds overwhelms our lives* (1st ed.). New York, NY: Metropolitan Books.

Goffman, E. (1966). *Behavior in public places: Notes on the social organization of gatherings*. New York, NY: Free Press.

Goffman, E. (1967). *Interaction ritual: Essays in face-to-face behavior*. Chicago, IL: Aldine Publishing Co.

Goldthorpe, J.E. (1996). *The sociology of post-colonial societies: Economic disparity, cultural diversity and development*. Cambridge; New York: Cambridge University Press.

Goody, J. (1977). *The domestication of the savage mind*. Cambridge; New York, NY: Cambridge University Press.

Goody, J. (2000). *The power of the written tradition*. Washington, DC: Smithsonian Institution Press.

Graff, H.J. (1979). *The literacy myth: Literacy and social structure in the nineteenth-century city*. New York, NY: Academic Press.

Grossman, E. (2006). *High tech trash: Digital devices, hidden toxics, and human health*. Washington, DC: Island Press.

Guha, R. (2000). *The unquiet woods : Ecological change and peasant resistance in the Himalaya* (Expanded ed.). Berkeley, CA: University of California Press.

Gupta, A. and Ferguson, J. (1997). *Anthropological locations: Boundaries and grounds of a field science*. Berkeley, CA: University of California Press.

Habermas, J. (1989). *The structural transformation of the public sphere*. Cambridge, MA: MIT Press.

Habibullah, W. and Ahuja, M. (2005). *Computerisation of land records*. New Delhi; Thousand Oaks, CA: Sage Publications.

Hafkin, N.J. and Huyer, S. (2006). *Cinderella or cyberella? Empowering women in the knowledge society*. Bloomfield, CT: Kumarian Press, Inc.

Hafkin, N.J. and Taggart, N. (2001). *Gender, information technology, and developing countries*. Washington, DC: United States Agency for International Development.

Harris, R.J. (1999). *A cognitive psychology of mass communication* (3rd ed.). Mahwah, NJ: Erlbaum.

Hart, C.H. (1993). *Children on playgrounds: Research perspectives and applications*. New York, NY: State Universiy of New York Press.

Hart, R.A. (1979). *Children's experience of place*. New York, NY: Irvington Publishing and Halsted Press.

Hartmann, B. (1995). *Reproductive rights and wrongs: The global politics of population control* (Rev. ed.). Boston, MA: South End Press.

Harvey, D. (2005). *A brief history of neoliberalism*. New York, NY: Oxford University Press.

Hauben, M. and Hauben, R. (1997). *Netizens: On the history and impact of usenet and the internet*. Los Alamitos, CA: IEEE Computer Society Press.

Hayes, N., Whitley, E.A., and Introna, L.D. (2006). *Power, knowledge and management information systems education: The case of the Indian learner.* Paper presented at the Twenty-Seventh International Conference on Information Systems, March 8, 2006, Milwaukee, WI.

Held, D. and McGrew, A.G. (2007). *Globalization theory: Approaches and controversies*. Cambridge: Polity Press.

Henwood, F. (2000). From the woman question in technology to the technology question in feminism. *European Journal of Women's Studies*, 7(2), pp. 209–27.

Hess, D.J. (1995). *Science and technology in a multicultural world: The cultural politics of facts and artifacts*. New York, NY: Columbia University Press.

Hirst, P.Q. and Thompson, G. (1996). *Globalization in question: The international economy and the possibilities of governance*. Cambridge: Polity Press.

Hoeschele, W. (2000). Geographic information engineering and social ground truth in attappadi, kerala state, India. *Annals of the Association of American Geographers*, 90(2), pp. 293–321.

Hopkins, P.D. (1998). *Sex/machine: Readings in culture, gender, and technology*. Bloomington, IN: Indiana University Press.

Horst, H.A. and Miller, D. (2006). *The cell phone: An anthropology of communication*. Oxford: Berg.

Ikels, C. (2004). *Filial piety: Practice and discourse in contemporary East Asia*. Stanford, CT: Stanford University Press.

Inamdar, P. and Kulkarni, A. (2007). "Hole-in-the-wall" computer kiosks foster mathematics achievement – a comparative study. *Educational Technology & Society*, 10(2), pp. 170–79.

Institute for Financial Management and Research (IFMR). (2007). *Giving farmers direct entry to retail markets*. Retrieved on November 1, 2009 from India Development Blog: http://www.indiadevelopmentblog.com/2008/12/giving-farmers-direct-entry-to-markets.html.

James, J. (2003). *Bridging the global digital divide*. Cheltenham: Edward Elgar.

Jensen, R. (2007). The digital provide: Information (technology), market performance, and welfare in the south Indian fisheries sector. *Quarterly Journal of Economics*, 122(3), pp. 879–924.

Jodha, N.S. (2005). Adaptation strategies against growing environmental and social vulnerabilities in mountain areas. *Himalayan Journal of Sciences*, 3(5), pp. 33–42.

Kabeer, N. (1994). *Reversed realities: Gender hierarchies in development thought*. London: Verso.

Kabeer, N., Stark, A., and Magnus, E. (2008). *Global perspectives on gender equality: Reversing the gaze*. New York, NY: Routledge.

Keniston, K. and Kumar, D. (2004). *IT experience in India: Bridging the digital divide*. London: Sage Publications.

Kenway, J. (2006). *Haunting the knowledge economy*. London: Routledge.

Kress, G.R. and Van Leeuwen, T. (2006). *Reading images: The grammar of visual design* (2nd ed.). London: Routledge.

Labov, W. (1980). *Locating language in time and space*. New York, NY: Academic Press.

LaFollette, M.C. (1992). *Stealing into print: Fraud, plagiarism, and misconduct in scientific publishing*. Berkeley, CA: University of California Press.

Lanham, R.A. (1995). Digital literacy. *Scientific American*, 273(3), pp. 160–61.

Lankshear, C. and Knobel, M. (2005). *Digital literacies: Policy, pedagogy and research considerations for education*. Paper presented at the ITU Conference, April 23–24, Oslo, Norway.

Lathrop, A. and Foss, K. (2000). *Student cheating and plagiarism in the Internet era: A wake-up call*. Englewood, CA: Libraries Unlimited.

Latour, B. (1993). *We have never been modern*. New York, NY: Harvester Wheatsheaf.

Latour, B. (2005). *Reassembling the social: An introduction to actor-network-theory*. Oxford: Oxford University Press.

Lave, J. and Wenger, E. (1991). *Situated learning: Legitimate peripheral participation*. Cambridge: Cambridge University Press.

Lechner, F.J. and Boli, J. (2005). *World culture: Origins and consequences*. Malden, MA: Blackwell Publishing.

Lévi-Strauss, C. (1966). *The savage mind*. Chicago, IL: University of Chicago Press.

Mack, R.L. (2001). *The digital divide: Standing at the intersection of race and technology*. Durham, NC: Carolina Academic Press.

Malhan, I.V. and Rao, S. (2007). Agricultural knowledge transfer in India: A study of prevailing communication channels. *Library Philosophy and Practice*, 21(2), pp. 12–34.

Mankekar, P. (1993). National texts and gendered lives: An ethnography of television viewers in a north Indian city. *American Ethnologist*, 20(3), pp. 543–63.

Marcus, G.E. (1998). *Ethnography through thick and thin*. Princeton, NJ: Princeton University Press.

Marshall, S. (2003). *Closing the digital divide: Transforming regional economies and communities with information technology*. Westport: Praeger.

Maslow, A.H. (1954). *Motivation and personality* (1st ed.). New York, NY: Harper.

McLuhan, M. (1962). *The Gutenberg galaxy: The making of typographic man*. Toronto: University of Toronto Press.

McMichael, P. (2004). *Development and social change: A global perspective* (3rd ed.). Newbury Park, CA: Pine Forge Press.

Mehrotra, S.K. (2006). *The economics of elementary education in India: The challenge of public finance, private provision, and household costs*. New Delhi: Sage Publications.

Meyer, J.W., Boli, J., Thomas, G.M., and Ramirez, F.O. (1997). World society and the nation-state. *American Journal of Sociology*, 103(1), pp. 144–81.

Mgbeoji, I. (2006). *Global biopiracy: Patents, plants and indigenous knowledge*. Vancouver: University of British Columbia Press.

Mies, M. (1986). *Patriarchy and accumulation on a world scale: Women in the international division of labour*. London: Zed Books.

Miller, D. (2001). *Car cultures*. Oxford: Berg.

Miller, D. and Slater, D. (2000). *The Internet: An ethnographic approach*. Oxford: Berg.

Mitra, S. (2000). *Minimally invasive education for mass computer literacy*. Paper presented at the CRIDALA, June 21–25, 2000, Hong Kong.

Mitra, S. (2003). Minimally invasive education: A progress report on the "Hole-in-the-wall" Experiments. *British Journal of Educational Technology*, 34(3), pp. 367–71.

Mitra, S. (2004, September 23). The hole in the wall. *Dataquest*. Retrieved on November 1, 2009 from: http://dqindia.ciol.com/content/industrymarket/2004/104092301.asp.

Mitra, S. (2005). Self-organizing systems for mass computer literacy: Findings from the "Hole in the Wall" experiments. *International Journal of Development Issues*, 4(1), pp. 71–81.

Mitra, S. and Rana, V. (2002). Children and the internet: Experiments with minimally invasive education in India. *British Journal of Educational Technology*, 32(2), pp. 221–32.

Mitra, S., Dangwal, R., Chatterjee, S., and Jha, S. (2005). A model of how children acquire computing skills from hole-in-the-wall computers in public places. *Information Technologies and International Development Journal*, 2(4), pp. 41–60.

Mohanty, C.T. (2003). *Feminism without borders: Decolonizing theory, practicing solidarity*. Durham, NC: Duke University Press.

Mohanty, C.T., Russo, A., and Torres, L. (1991). *Third world women and the politics of feminism*. Bloomington, IN: Indiana University Press.

Moll, L.C. (1992). Bilingual classroom studies and community analysis: Some recent trends. *Educational Researcher*, 21(2), pp. 20–24.

Monroe, B.J. (2004). *Crossing the digital divide: Race, writing, and technology in the classroom*. New York, NY: Teachers College Press.

Narasimhan, S. (1999). *Empowering women: An alternative strategy from rural India*. New Delhi: Sage Publications.

Narayan, D. (2000). *Voices of the poor: Can anyone hear us?* Oxford: Oxford University Press.

National Informatics Centre (NIH). (2005). *Good governance through ICT*. Retrieved on November 1, 2009 from: www.nic.in/nicportal/DocumentsPDFs/ICTBook.pdf.

Negi, S.S. (1995). *Uttarakhand: Land and people*. New Delhi: MD Publications.

Negroponte, N. (1995). *Being digital* (1st ed.). New York, NY: Knopf.

Norman, D.A. (1990). *The design of everyday things* (1st Doubleday/Currency ed.). New York, NY: Doubleday.

Norris, P. (2001). *Digital divide: Civic engagement, information poverty, and the Internet worldwide.* Cambridge: Cambridge University Press.

O'Donnell, A.M. and King, A. (1999). *Cognitive perspectives on peer learning.* Mahwah, NJ: Erlbaum.

OECD. (2006). *India: E-readiness assessment report 2006.* Retrieved on October 29, 2009 from: http://74.125.47.132/search?q=cache:6DerWQCYgKQJ:www. mit.gov.in/download/eready2006/Forewardcontents.PDF+India:+e-Readiness +Assessment+Report+2006&hl=en&ct=clnk&cd=1&gl=us.

Olson, D.R. and Torrance, N. (1991). *Literacy and orality.* Cambridge: Cambridge University Press.

Ong, W.J. (1988). *Orality and literacy: The technologizing of the word.* London: Routledge.

Opie, I.A. and Opie, P. (1969). *Children's games.* New York, NY: Oxford University Press.

Osborne, S.P. (2000). *Public-private partnerships: Theory and practice in international perspective.* London: Routledge.

Oudshoorn, N. and Pinch, T.J. (2003). *How users matter: The co-construction of users and technologies.* Cambridge, MA: MIT Press.

Panagariya, A. (2008). *India: The emerging giant.* Oxford: Oxford University Press.

Pathak, R.D. and Prasad, S.R. (2006). *Role of e-governance in tackling corruption: The Indian experience.* Delhi: Center for Media Studies.

Peet, R. and Hartwick, E.R. (1999). *Theories of development.* New York, NY: Guilford Press.

Pellegrini, A.D. (1995). *School recess and playground behavior: Educational and developmental roles.* Albany, NY: State University of New York Press.

Pennycook, A. (1996). *Borrowing others' words: Text, ownership, memory, and plagiarism.* New York, NY: TESOL Ltd.

Perraton, H.D. (2000). *Open and distance learning in the developing world.* London: Routledge.

Phillipson, R. (1992). *Linguistic imperialism.* Oxford: Oxford University Press.

Prahalad, C.K. (2005). *The fortune at the bottom of the pyramid: Eradicating poverty through profits: Enabling dignity and choice through markets.* Upper Saddle River, NJ: Wharton School Publishing.

Prakash, O. (2007). *ICT in Agriculture.* Presentation by Secretary of Agriculture at eIndia 2009, Hyderabad, India. Retrieved on November 1, 2009 from: http:// www.eindia.net.in/eagriculture/presentation/Om_Prakash.pdf.

Prensky, M. (2001). Digital natives, digital immigrants. *On the Horizon,* 9(5), pp. 1–6.

Putler, D.S. and Zilberman, D. (1988). Computer use in agriculture: Evidence from Tulare county, California. *American Agricultural Economics Association,* 70(4), pp. 790–802.

Putnam, R.D. (2002). *Democracies in flux: The evolution of social capital in contemporary society*. Oxford: Oxford University Press.

Rangaswamy, N. and Toyoma, K. (2006). *Global events local impacts: India's rural emerging markets*. Paper presented at the Ethnographic Praxis in Industry Conference, Portland, Oregan.

Redfield, R. (1955). *The little community: Viewpoints for the study of a human whole*. Chicago, IL: University of Chicago Press.

Rheingold, H. (2000). *The virtual community: Homesteading on the electronic frontier* (Rev. ed.). Cambridge, MA: MIT Press.

Rheingold, H. (2003). *Smart mobs: The next social revolution*. Cambridge, MA: Basic Books.

Rheingold, H. and Kluitenberg, E. (2006). Mindful disconnection: Counterpowering the panopticon from the inside. *Hybrid Space*, 11(2), pp. 1–12.

Roberts, K. (2006). *Leisure in contemporary society* (2nd ed.). Wallingford: CABI.

Russell, B. (1960). *In praise of idleness and other essays*. London: George Allen and Unwin.

Sachs, J. (2005). *The end of poverty: Economic possibilities for our time*. New York, NY: Penguin Press.

Samanta, R.K. (1999). *Empowering women: Key to third world development*. New Delhi: MD Publications.

Samoff, J. (1974). *Tanzania: Local politics and the structure of power*. Madison, WI: University of Wisconsin Press.

Samoff, J. (1995). *Analyses, agendas and priorities in African education: A review of externally initiated, commissioned and supported studies of education in Africa, 1990–1994* (Rev. ed.). Paris: UNESCO.

Sanjek, R. and Colen, S. (1990). *At work in homes: Household workers in world perspective*. Washington, DC: American Anthropological Association.

Sassen, S. (2006). *Cities in a world economy* (3rd ed.). Thousand Oaks, CA: Pine Forge Press.

Sassen, S. (2006a). *Territory, authority, rights: From medieval to global assemblages*. Princeton, NJ: Princeton University Press.

Sati, M.C. and Sati, S.P. (2000). *Uttarakhand statehood: Dimensions of development*. New Delhi: Indus Publishing Co.

Satyanarayana, K.V. and Babu, R.B. (2008). *Trends in the development of e-theses in India: Issues, constraints, and solutions*. Pune: Tata Research Development and Design Centre, Tata Consultancy Services Ltd.

Scott, W.R. (1995). *Institutions and organizations*. Newbury Park, CA: Sage Publications.

Sen, A.K. (1999). *Development as freedom*. Oxford: Oxford University Press.

Sen, A.K., Basu, K., Pattanaik, P.K., and Suzumura, K. (1995). *Choice, welfare, and development*. Oxford: Oxford University Press.

Sethi, R.M. (1999). *Globalization, culture, and women's development*. Jaipur: Rawat Publications.

Shiva, V. (1991a). *Staying alive: Women, ecology, and development*. London; Atlantic Highlands, NJ: Zed Books.

Shiva, V. (1991b). *The violence of the green revolution: Third world agriculture, ecology, and politics*. London: Zed Books.

Sillitoe, P. (2000). *Indigenous knowledge development in bangladesh: Present and future*. Florida: University Press.

Singhal, A. and Rogers, E.M. (1989). *India's information revolution*. New Delhi: Sage Publications.

Sivaramakrishnan, K. and Agrawal, A. (2003). *Regional modernities: The cultural politics of development in India*. Oxford: Oxford University Press.

Solomon, G., Allen, N.J., and Resta, P.E. (2003). *Toward digital equity: Bridging the divide in education*. Boston, MA: Allyn and Bacon.

Sreekumar, T.T. (2007). Cyber kiosks and dilemmas of social inclusion in rural India. *Media Culture & Society*, 29(6), pp. 869–89.

Stebbins, R.A. (1992). *Amateurs, professionals and serious leisure*. Montreal: McGill-Queen's University Press.

Stokes, E. (1989). *The English utilitarians and India*. New York, NY: Oxford University Press.

Street, B.V. (1993). *Cross-cultural approaches to literacy*. Cambridge, MA; New York: Cambridge University Press.

Sutton-Smith, B. (1979). *Play and learning*. New York, NY: Gardner Press.

Sutton-Smith, B. (1997). *The ambiguity of play*. Cambridge: Harvard University Press.

Tedlock, D. and Mannheim, B. (1995). *The dialogic emergence of culture*. Urbana, IL: University of Illinois Press.

Thomas, L.S. and Salvador, T. (2006). *Skillful strategy, artful navigation and necessary wrangling*. Paper presented at the Ethnographic Praxis in Industry Conference, September 24–26, 2006, Portland, Oregon.

Thurlow, C., Lengel, L.B., and Tomic, A. (2004). *Computer mediated communication: Social interaction and the Internet*. London: Sage Publications.

Tiffin, J. and Rajasingham, L. (1995). *In search of the virtual class: Education in an information society*. London: Routledge.

Trivedi, V.R. (1995). *Autonomy of Uttarakhand*. New Delhi: Mohit Publications.

Tudge, J. and Winterhoff, P. (2006). Can young children benefit from collaborative problem solving? Tracing the effects of partner competence and feedback. *Social Development*, 2(3), pp. 242–59.

Turkle, S. (1984). *The second self: Computers and the human spirit*. New York, NY: Simon & Schuster.

Turkle, S. (1995). *Life on the screen: Identity in the age of the Internet*. New York, NY: Simon & Schuster.

Tyner, K.R. (1998). *Literacy in a digital world: Teaching and learning in the age of information*. Mahwah, NJ: Erlbaum.

United Nations Development Programme. (2009). Poverty reduction fact sheets. Retrieved on November 1, 2009 from: http://www.undp.org.in/index.php?option=com_content&view=article&id=220&Itemid=580.

Varenne, H. and McDermott, R. (1998). *Successful failure*. Jackson, TN: Westview Press.

Varma, R. (2002). Technological fix: Sex determination in India. *Science, Technology & Society*, 22(1), pp. 21–30.

Villereal, G.L. (2007). Guatemala's current and future globalization. *International Social Work*, 50(1), pp. 41–51.

Vygotsky, L. (1978). *Mind and society: The development of higher mental processes*. Cambridge, MA: Harvard University Press.

Wagner, D.A. (1999). *The future of literacy in a changing world* (Rev. ed.). Cresskill, NJ: Hampton Press.

Wajcman, J. (1991). *Feminism confronts technology*. Cambridge: Polity Press.

Waldman, A. (2004). India's soybean farmers join the global village. *The New York Times*, January 1.

Warschauer, M. (2003). *Technology and social inclusion: Rethinking the digital divide*. Cambridge, MA: MIT Press.

Wignaraja, P. and Sirivardana, S. (2004). *Pro-poor growth and governance in South Asia: Decentralization and participatory development*. New Delhi: Sage Publications.

Willis, P.E. (1981). *Learning to labor: How working class kids get working class jobs*. New York, NY: Columbia University Press.

Zinn, H. (2003). *A people's history of the United States: 1942–present*. New York, NY: HarperCollins.

Zivin, J. (1998). The imagined reign of the iron lecturer: Village broadcasting in colonial India. *Modern Asian Studies*, 32(3), pp. 717–38.

Index